ADHESION 12

This volume is based on papers presented at the 25th annual conference on Adhesion and Adhesives held at The City University, London

Previous conferences have been published under the titles of
Adhesion 1–11

ADHESION 12

Edited by

K. W. ALLEN

Adhesion Science Group,
The City University, London, UK

ELSEVIER APPLIED SCIENCE
LONDON and NEW YORK

ELSEVIER APPLIED SCIENCE PUBLISHERS LTD
Crown House, Linton Road, Barking, Essex IG11 8JU, England

Sole Distributor in the USA and Canada
ELSEVIER SCIENCE PUBLISHING CO., INC.
52 Vanderbilt Avenue, New York, NY 10017, USA

WITH 39 TABLES AND 140 ILLUSTRATIONS

© ELSEVIER APPLIED SCIENCE PUBLISHERS LTD 1988

© PERMABOND—Chapter 9
Softcover reprint of the hardcover 1st edition 1988

British Library Cataloguing in Publication Data

Conference on Adhesion and Adhesives *(25th:*
1987: City University)
Adhesion 12.
1. Adhesion
I. Title II. Allen, K. W.
541.3′453

ISBN-13: 978-94-010-7100-0   e-ISBN-13: 978-94-009-1349-3
DOI: 10.1007/978-94-009-1349-3
Library of Congress Cataloging in Publication Data

Conference on Adhesion and Adhesives (25th: 1987: City University,
London)
Adhesion 12/edited by K. W. Allen.
  p. cm.
"This volume is based on papers presented at the 25th Annual
Conference on Adhesion and Adhesives held at the City University,
London"—P. opposite t.p.
  Includes bibliographies and index.

  1. Adhesion—Congresses. 2. Adhesives—Congresses. I. Allen, K.
W. II. Title. III. Title: Adhesion twelve.
TP967.C64 1987
668′.3—dc 19   88–308
                                                     CIP

Special regulations for readers in the USA

Preface

Twenty-five years is a considerable time span in the life of any scientific discipline; certainly in this twentieth century when development is so rapid. For the science of adhesion and the technology of adhesives this is particularly true. For these, the immediately past quarter century might be compared with the Renaissance when all the civilised world was alight with the 'new learning'. Certainly it has been a period of immense advance both of understanding and of application in this area of scientific endeavour.

It was in the light of this situation that here at City University we set about arranging the Twenty-fifth Annual Conference on Adhesion and Adhesives, of which this volume presents the proceedings. A total of seventeen papers from seven countries, covering a span of topics from organic chemistry through physical chemistry and physics to engineering. Truly this Conference has 'come of age' and is acknowledged as the annual international venue for the consideration of adhesion in all its diversity. It is our earnest hope and intention that it shall continue for many more years. May I express my personal gratitude to all those who make the event possible; the audiences as well as the speakers, all those in the University who help in various ways, and the publishers who make it possible for you, the wider audience, to have these proceedings. There is one person of whom I must make special mention for thanks both in connection with these Conferences and in so many other ways and that is Dr 'Bill' Wake the doyen of British Adhesion Science. To you all, 'Thank you very much' not only for this year but for all the years past as well.

K. W. ALLEN

Contents

Preface v

1. *The Influence of the Surface State of Polymers on the Determination of the Contact Angle* 1
   C. Bischof, R.-D. Schulze and W. Possart (Academy of Sciences of the GDR, Institute of Polymer Chemistry 'Erich Correns', Kantstrasse 55, Teltow 1530, GDR)

2. *Factors Affecting the Durability of Titanium/Epoxy Bonds* . 17
   J. A. Filbey and J. P. Wightman (Department of Chemistry, Center for Adhesion Science, Virginia Polytechnic Institute & State University, Blacksburg, VA 24061, USA)

3. *The Role of Adhesion in Mineral Filled Plastics* . . . 33
   P. J. Wright (Blue Circle Industries PLC, Greenhithe, Kent, UK)

4. *Adhesion and Protective Properties of Polyamidoimide Coatings on Copper Substrate* 55
   V. M. Startsev, R. Z. Kaz'mina, V. A. Ogarev and V. N. Buryanenko (Department of Polymer Coatings, Academy of Sciences of the USSR, Lininsky prospekt, 31, Moscow 117915, USSR)

5. *Adhesion of Nitrile and Ethylene-Propylene Rubber to Mould Materials* 69
   R. K. Champaneria, M. Lotfipour and D. E. Packham (School of Materials Science, University of Bath, Bath BA2 7AY, UK)

6. *Improvement and Diversification of Cyanoacrylate Adhesives* . 82
   D. L. Kotzev and V. S. Kabaivanov (Scientific Industrial
   Centre for Special Polymers, Kliment Ohridski St. 4A, 1156
   Sofia, Bulgaria)

7. *Polymerisation and Surface Interaction of a Silane Coupling
   Agent* 106
   P. J. Grey (BTR Industries Ltd, Central Development,
   Burton-on-Trent, Staffs DE13 0SN, UK)

8. *Applications of Toughened Epoxy and Acrylic Adhesives* . 121
   K. W. Harrison (Bostik Ltd, Leicester LE4 6BW, UK)

9. *Stress in Bonded Joints* 141
   W. A. Lees (Permabond Adhesives Ltd, Woodside Road,
   Eastleigh, Hampshire SO5 4EX, UK)

10. *An Adhesive Interleaf to Reduce Stress Concentrations Between
    Plys of Structural Composites* 159
    R. B. Krieger Jr (American Cyanamid Co., Engineered
    Materials Department, Old Post Road, Havre de Grace, MD
    21078, USA)

11. *Chemistry of Phenolic and Epoxy Adhesives* 166
    H. Kollek (Fraunhofer-Institut fur Angewandte Material-
    forschung, Lesumer Hierstrasse 36, D-2820 Bremen 77
    (Lesum), FRG)

12. *The Effect of the Adhesive Thickness on the Strength of a Bonded
    Joint* 174
    A. D. Crocombe (Department of Mechanical Engineering,
    University of Surrey, Guildford, Surrey GU2 5XH, UK)

13. *Bonding of Composites to Metals* 193
    R. D. Adams (Department of Mechanical Engineering, Uni-
    versity of Bristol, Bristol BS8 1TR, UK) and J. A. Harris
    (Materials Engineering Research Laboratory, Tamworth
    Road, Hertford SG13 7DG, UK)

14. *Stick-Slip and Peeling of Adhesive Tapes* 205
    D. Maugis and M. Barquins (Laboratoire de Mecanique des
    Surfaces du CNRS LCPC, 58 Bd Lefebvre, 75732 Paris
    Cedex 15, France)

15. *Hard Rubber/Metal Adhesion Assessment Using a Heavy
    Cylinder Rolling Test* 223
    M. E. R. Shanahan, N. Zaghzi, J. Schultz (Centre de
    Recherches sur la Physico-Chimie des Surfaces Solides and
    Ecole Nationale Superieure de Chimie de Mulhouse, 3 Rue
    Alfred Werner, 68093 Mulhouse Cedex, France) and A.
    Carré (Corning Europe Inc., 77211 Avon Cedex, France)

16. *A Modified Thermoplastic Adhesive Layer in Laminates* . 239
    R. J. Ashley (Product Innovation Section, Metal Box plc,
    R & D Division, Wantage, Oxon OX12 9BP, UK)

17. *The Use of Electron Microscopy for the Analysis of the Adhesive–
    Adherend Interface in the Aluminium–Aluminium Bonded Joint* 248
    J. A. Bishopp (Bonded Structures, Ciba-Geigy Plastics,
    Duxford, Cambridge CB2 4QD, UK) and E. K. Sim, T. V.
    Smith, G. E. Thompson and G. C. Wood (Corrosion and
    Protection Centre, University of Manchester, Institute of
    Science and Technology, Manchester M60 10D, UK)

THE INFLUENCE OF THE SURFACE STATE OF POLYMERS ON THE DETERMINATION OF THE CONTACT ANGLE

C. Bischof, R.-D. Schulze and W. Possart
Academy of Sciences of the G.D.R.
Institute of Polymer Chemistry "Erich Correns"
Kantstraße 55
Teltow 1530, G.D.R.

and

H. Kamusewitz
Padagogische Hochschule
"N.K. Krupskaja" Halle
G.D.R.

1. INTRODUCTION

It was Gibbs who developed the thermodynamics of phase boundaries to a well established theory. To a great part, his success resulted from neglecting the details of the molecular structure and of the particle dynamics, as is common for a thermodynamic treatment. This not only simplifies the interpretation of the phenomena at the phase boundary; additionally the measurements reduce to the determination of a few thermodynamic quantities. On the other hand, the thermodynamic point of view limits the possible information and conclusions about the state of thermal equilibrium of the phases considered, including their interface layers. The discussion of structure or molecular motions requires the use of additional models.

These facts explain the fast development of the thermodynamic adhesion theories compared with other attempts. We emphasize the important contributions of Zisman, Fowkes and Good, although many other scientists have reported valuable results.

As a beginning we remember a few thermodynamic terms and relations. The boundary layer between the phases a and b is described by the interfacial tension $\gamma_{a,b}$, confirming a tangent onto the interface. According to Bakker's equation, the absolute value of the interfacial tension embodies an integral feature of the boundary layer and it equals the specific free energy of the Gibbsian interface. For the tension of a phase against air (or vapour) the term "surface tension" is commonly used but it has the same meaning as the interfacial tension just introduced.

Consider now two separate phases a and b both in contact with a third phase c (e.g. air) and so possessing the specific free interface energies $\gamma_{a,c}$ and $\gamma_{b,c}$. A contact between a and b forms a new interface of the free energy $\gamma_{a,b}$. This process obeys the energy balance

$$\gamma_{a,c} + \gamma_{b,c} = \gamma_{a,b} + W_A \tag{1}$$

where W_A is the energy gain called by Dupré "the specific <u>thermodynamic work of adhesion</u>". Hence this W_A is defined for a <u>reversible</u> creation of the new boundary layer a – b and it is by no means equal to the work of separation spent in mechanical testing.

It follows from the thermodynamic context that W_A corresponds to the change in the specific free interfacial energy of the system during the creation of the new layer out of a part of the former layers with $W_A \geqslant 0$.

Returning to the interface tension we remember for the triple line the thermodynamically based relation

$$\bar{\gamma}_{a,b} + \bar{\gamma}_{a,c} + \bar{\gamma}_{b,c} = 0 \tag{2}$$

which changes for the system solid – liquid – vapour into the well-known Young's equation

$$\gamma_{s,v} - \gamma_{s,l} = \gamma_{l,v} \cdot \cos \theta_e \tag{3}$$

as it is illustrated in Fig. 1. The contact angle θ_e refers to thermal equilibrium as it is indicated by the subscript e [1,2]. Inserting into eq. (1) yields the specific thermodynamic work of adhesion

$$W_A = \gamma_{1,v} \, (1 + \cos \theta_e) \tag{4}$$

The importance of eq. (4) is due to the fact that it connects W_A to measurable quantities only.

Whilst the surface tension $\gamma_{1,v}$ of liquids can be measured by various methods with good accuracy, for contact angle measurements a hysteresis $\Delta\theta$ between advancing angle θ_a and receding angle θ_r is observed on nearly every real solid surface. Up to now, the answer to the question, how the thermodynamic equilibrium contact angle θ_e should be deduced from the measured quantities θ_a and θ_r, is not completely convincing. The Sections 2 and 3 attempt to arrange and extend the common concepts to some degree. Further, an empirical concept for the determination of θ_e is presented.

For now we should keep in mind that the adhesion scientist is more interested in the work of adhesion W_A for two solid phases s1 and s2. In this case eq. (1) changes into

$$W_A = \gamma_{s1,v} + \gamma_{s2,v} - \gamma_{s1,s2} \tag{5}$$

and we should like to know all the three interface energies. In order to solve this, an additional relation between the three energies is needed. Among the numerous variants proposed in the literature we prefer the equation

$$\gamma_{s1,s2} = \gamma_{s1,v} + \gamma_{s2,v} - 2 \, \Phi_{s1,s2} \, \sqrt{\gamma_{s1,v} \cdot \gamma_{s2,v}} \tag{6}$$

which was developed by Good [3-5]. This equation generalizes some former ideas. In our opinion it is very closely related to the thermodynamics of the phase boundary and Good gives a clear and elegant interpretation of the interaction parameter Φ on the basis of molecular interaction theory.

Inserting eq. (6) into (5) provides

$$W_A = 2. \, \Phi_{s1,s2} \cdot \sqrt{\gamma_{s1,v} \cdot \gamma_{s2,v}} \tag{7}$$

Besides the interaction parameter $\Phi_{s1,s2}$ for the interface between the two solids, their specific free surface energies must be known. These quantities are estimated from the determination of the equilibrium contact angle θ_e which pure simple liquids form on these solids. For this experiment, the Good equation (6) takes the form

$$\gamma_{s,1} = \gamma_{s,v} + \gamma_{1,v} - 2 \phi_{s,1} \sqrt{\gamma_{s,v} \cdot \gamma_{1,v}} \tag{8}$$

and with the Young's equation (3) the surface tension of the solid is given by:

$$\gamma_{s,v} = \gamma_{1,v} \frac{(1 + \cos \theta_e)^2}{4 \phi_{s,1}^2} \tag{9}$$

Within their frame of applicability these equations (7 and 9) provide the specific thermodynamic work of adhesion for solid-solid contacts. Note that this quantity summarizes the effect of all kinds of intermolecular interaction excepting chemical bonds.

It should be emphazised that the two interaction parameters involved in equations (7 and 9) are not easily determined. In view of the complete lack of measuring methods, only theoretical estimations are available. Here, advances would be especially welcome in the future. In Section 4 we present a way in which $\gamma_{s,v}$ could be estimated directly from contact angle data for simple liquids.

2. AN ATTEMPT TO INTERPRET THE CONNECTION BETWEEN θ_e AND CONTACT ANGLE MEASUREMENT

2.1 Basic Considerations

According to the thermodynamic theory, in the three-phase system solid - liquid - vapour (s - 1 - v), the liquid forms a meniscus which forms with the solid surface the characteristic contact angle θ_e at the triple line D (see Fig. 1). This angle is correlated via eq. (3) with the interfacial tensions of the three phases. The shape of the whole liquid meniscus follows as a solution of the Laplace equation

$$p_h = p_{1,v} = \gamma_{1,v} \quad (k_1 + k_2) \tag{10}$$

which is valid for the equilibrium between hydrostatic pressure p_h and interface pressure $p_{1,v}$. The local main curvatures of the meniscus are designed by k_1 and k_2.

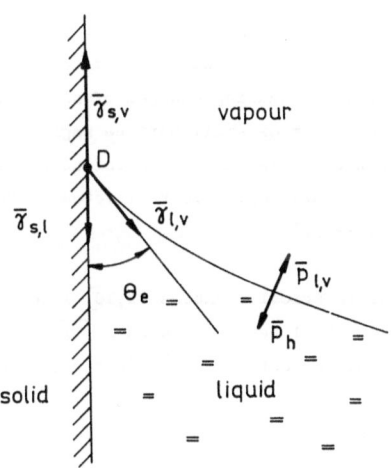

Figure 1   Forces on meniscus in solid-liquid-vapour systems

In the experiment, the tangent is put on the plane projection of the curved meniscus (Fig. 1). But instead of θ_e, the advancing angle θ_a and the receding angle θ_r are found after enlarging or reducing the area of liquid-solid contact. This contact angle hysteresis $\Delta\theta = \theta_a - \theta_r$, exists although the whole system is in its equilibrium. On roughened surfaces a hysteresis up to 90^0 is observed [6] and even on highly polished surfaces it amounts to a few degrees [7].

2.2   Causes of Contact Angle Hysteresis

The contact angle hysteresis may be caused by temporary and local fluctuations of the system as well.

Time-dependent changes in the system counterfeit an apparent hysteresis since θ_a and θ_r are measured subsequently. All processes modifying the interfacial tensions belong to this category (e.g. chemical reactions, swelling, adsorption, reorientation of molecules, composition changes of the boundary layer by contamination or demixing, and the like). They are excluded from the further discussion since they invalidate the presumption of thermal equilibrium. In the experiment such effects are easily recognizable by repeated measurements of θ_a and θ_r.

Local fluctuations are mainly confined to the solid surface as far as low molecular pure liquids are used. The fluctuations occur as inhomogeneities of composition causing changes of solid surface tension or as a varying spatial orientation of small surface segments (roughness). At present, different opinions exist about the influence of the two kinds of fluctuation. Mostly, the discussion is restricted to the qualitative features (compare e.g. [8]).

Besides various empirical formulae [9,10] like $\theta_e = \theta_a$ or $\theta_e = \frac{1}{2}(\theta_a + \theta_r)$ used in the literature; Wenzel [11] proposed a relation between θ_e and the relative surface enlargement r caused by the roughness

$$\cos \theta_m = r \cdot \cos \theta_e \tag{11}$$

It remains unclear however which of the two measured angles has to be taken for θ_m. Other authors utilize idealized models of the surface shape in order to study the influence of roughness (e.g. [6]).

Now we present a new relation between θ_e and the hysteresis $\Delta\theta$. The derivation starts with a statisically rough solid surface and therefore is not limited to a special geometry. The influence of inhomogeneities of the interfacial tensions of the solid ($\gamma_{s,v}$; $\gamma_{s,l}$) is mentioned. Details are to be published elsewhere.

2.3 General Influences of the Roughness on the Measurable Contact Angles

The discussion starts with the basic formulae given in Section 2.1. In the vicinity of the triple line, the hydrostatic pressure p_h in eq. (10) may be considered as constant to a good approximation. Hence the sum of the main curvatures of this meniscus region possesses a constant value too. Figure 2 depicts, as an example, the meniscus at a vertical wall with vertical grooves. Since the triple line D follows the contour of the rough surface, each of the two main curvatures varies from point to point. These local fluctuations of the curvatures compensate at a small distance x from the solid surface. Additionally, an observer measures the angle θ_m on the projection at this distance x from the surface since he cannot look into the grooves. For small x, the observed angle θ_m is approximately equal to the virtual contact angle θ_a formed by the smoothed meniscus S_a (Fig. 3).

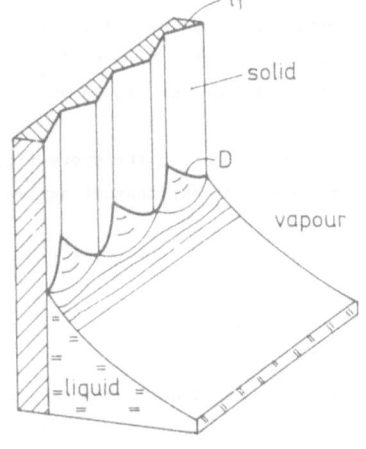

Figure 2
Forces on meniscus
in solid-liquid-vapour
systems

Figure 3
Section drawing of
meniscus on rough
surface as in Fig.2.

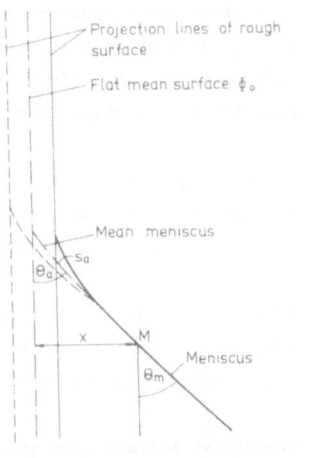

Detailed discussion [12] provides the equation

$$\cos \theta_m = K \cos (\theta_e + \delta) \tag{12}$$

The parameter K is given as the quotient of the length of the triple line D in its projection perpendicular to the solid surface and of the virtual length of this line as it is measured by an observer who is not able to resolve the surface profile. The second parameter results from a special averaging procedure over the inclinations of the microscopic surface segments along the triple line. Both parameters are influenced by the roughness and by the actual shape of the triple line which in turn depends on whether the liquid advances or recedes.

Thus, we obtain for every liquid-solid system with defined roughness, two equations for the contact angles

$$\cos \theta_a = K_a \cos (\theta_e + \delta_a) \tag{13.1}$$

$$\cos \theta_r = K_r \cos (\theta_e + \delta_r) \tag{13.2}$$

and the following conclusions can be drawn:

- The hysteresis measured for a defined solid-liquid-vapour system depends on the roughness via four parameters where K_a, δ_a and K_r, δ_r are pairwise interrelated. Additionally, the parameters are influenced by the surface tension of the liquid since shape and position of the triple line changes with θ_e.

- For an actual system, defined by θ_e, θ_a and θ_r, the equations (13) contain two additional degrees of freedom. Therefore, different roughness geometries may provide identical experimental values.

- Although the eqs. (13) do not allow for a straightforward determination of θ_e from a single measurement, they describe the influence of all possible kinds of roughness.

- Finally it follows from our model that the contact angle hysteresis is not determined by the depth but by the shape of the roughness.

This is confirmed by experiments [13].

Note that consideration of flat inhomogeneous solids yields equations of the same structure as eqs. (13) but with $K_a = K_r = 1$. Consequently, fluctuations of the solid surface tension produce the same type of hysteresis as roughness does.

2.4 Description of Model Geometries

Firstly, we consider the identical vertical grooves shown in Fig. 2. The triple line moves along the grooves. Obviously we have $K_a = K_r > 1$ and $\delta_a = \delta_r = 0$. This surface produces no hysteresis but note that $\theta_m \neq \theta_e$. Formerly eq. (13) reduces to the Wenzel equation (11) up to the different meaning of K and r. The parameter K reflects the effect of so-called capillarity, causing an enhanced spreading for $\theta_e < 90^0$ or an increased dewetting for $\theta_e > 90^0$.

For a second example, the grooves are horizontally arranged on the vertical wall. Now the triple line moves perpendicularly to the grooves. For a slow, continuous rise of the liquid, the triple line only resides on those side-planes of the grooves which are turned away from the liquid surface. The triple line jumps across all the side-planes turned towards the liquid surface since the meniscus shape on these planes would correspond to a sudden rise of the liquid surface. Comparable considerations follow for the receding triple line. As the result we obtain the so-called barrier effect. For α as the angle of inclination of side-planes of the groove we have

$$\delta_a = \alpha \text{ and } \delta_r = -\alpha, \quad \text{with } K_a = K_r = 1.$$

Obviously the barrier effect produces a hysteresis $\Delta\theta = 2\alpha$ and the eqs. (13) reduce for this special geometry to

$$\theta_e = \frac{1}{2} (\theta_a + \theta_r).$$

3. AN EMPIRICAL RELATION BETWEEN THERMODYNAMIC CONTACT ANGLE θ_e AND HYSTERESIS $\Delta\theta$ FOR POLYMERS

As a result of extensive investigations of various polymers we could show that the advancing and the receding angle are reproduced by several experimental methods (pendant and sessile drop, Wilhelmy plate) [14]. The measurements were carried out on PTFE, PE, EVAl-copolymer, PMMA and PS of varying roughness with the low-molecular liquids water, glycerol, formamide, ethylene-glycol, n-hexadecane and α-bromo-naphthalene. Figure 4 shows the empirical relations between the angles obtained θ_a, θ_r and the hysteresis $\Delta\theta$ as a measure of roughness. The following general features are established:

- Both functions $\theta_a = f(\Delta\theta)$ and $\theta_r = f(\Delta\theta)$ are well described by straight lines showing an intersection at $\Delta\theta = 0$. For $\Delta\theta$ as a measure of roughness, the extrapolation to $\Delta\theta = 0$ corresponds to an absolute flat surface and we get for the intersection on the ordinate the thermodynamic contact angle θ_e (for paraffin-water system see Fig. 4):

$$\theta_a = \theta_e + a \cdot \Delta\theta$$

$$\theta_r = \theta_e + b \cdot \Delta\theta; \quad a,b = \text{constants} \tag{14}$$

- The dependence of the mean value $\bar{\theta} = \frac{1}{2}(\theta_a + \theta_r)$ on $\Delta\theta$ obeys the function

$$\bar{\theta} = \theta_e + B \cdot \Delta\theta \qquad \text{with } B = \frac{a + b}{2} \tag{15}$$

where the slope B is a characteristic parameter for each polymer-liquid pair (for PTFE see Fig. 5). Plotting the values of B for all systems under investigation against the extrapolated θ_e we find a linear approximation (Fig. 6)

$$B = -1.096 + 0.0147 \cdot \theta_e \tag{16}$$

By inserting eq. (16) into eq. (15) we obtain

$$\theta_e = \frac{\theta + 1.086 \cdot \Delta\theta}{1 + 0.0147 \cdot \Delta\theta} \tag{17}$$

with all angles given in degrees. This empirical equation permits

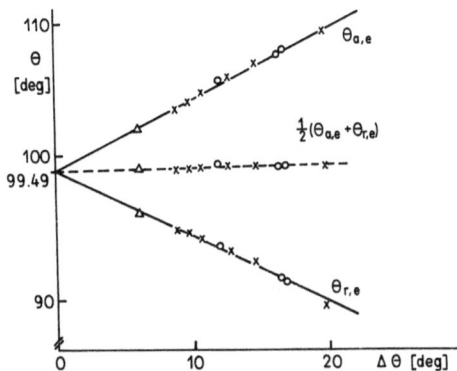

Figure 4. Contact angles as a function of hysteresis
paraffin-water-vapour system with different
solid surface roughnesses: melt casted on
glass, x grooves // to drop spreading, O
grooves ⊥ to drop spreading; coefficient of
linear regression R = 0.994.

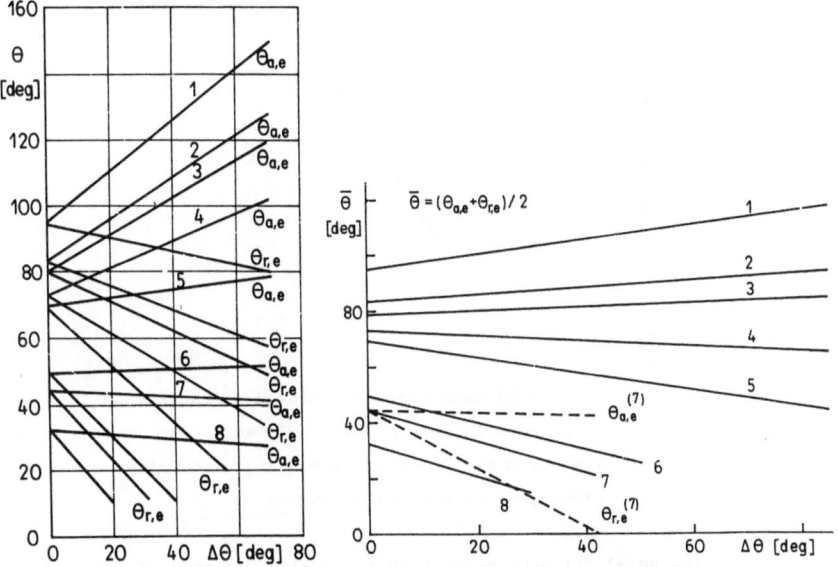

Figure 5: Contact angles as a function of hysteresis on
rough PTFE surfaces: linear approximation for
low molecular liquids: 1 water, 2 glycerol,
3 formamide, 4 ethylene glycol, 5 α-bromo-
naphthaline, 6 dioctylphthalate, 7 n-hexadecane and 8 n-decane.

the calculation of θ_e from a single measurement without any roughness variation. The validity of this result is tested in Section 4 for the polymer-liquid systems considered here.

In view of the theoretical consideration in Section 2 we point out that the eqs. (14,15) should be correct for a certain class of stochastically rough surfaces only. For this class, the hysteresis $\Delta\theta$ provides an integral measure of roughness. A first mathematical interdependence between the general parameters of eqs. (13) and $\Delta\theta$ may be found by substituting θ_e in eq. (13) with the help of eq. (17).

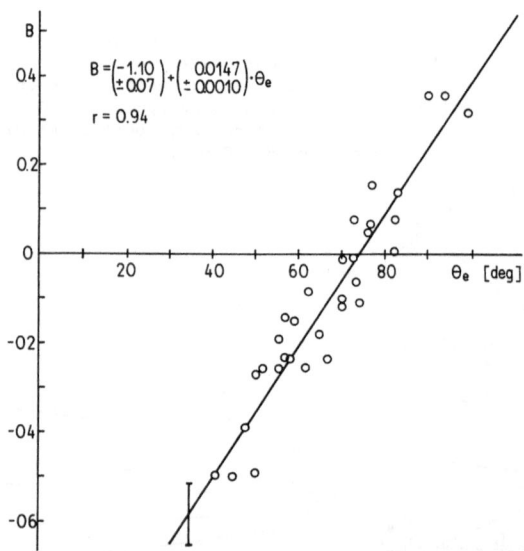

Figure 6: Slope B of mean contact angle as a function of hysteresis against extrapolated θ_e for systems under investigation O experimental slopes; — linear approximation.

The two model surfaces considered in Section 2.4 serve as examples for roughness types not belonging to the class found in our experiments. As mentioned above, the vertical grooves show no hysteresis and $\bar{\theta}$ is not equal to θ_e as it would be expected from (17) for $\Delta\theta = 0$. Horizontal grooves produce $\Delta\theta = 2\alpha$ and $\theta_e = \frac{1}{2}(\theta_a + \theta_r)$ according to eqs. (13). Hence this θ_e value again differs from that what would be calculated from eq.(17). Consequently, the application of the empirical relation (17) to oriented surface profiles requires a test if different planes of meniscus projection result in a change of θ_e. Concerning this point, our measurements provided no significant variations although some of the surfaces showed an apparent orientation of their surface geometry. Obviously not all we can see acts on the triple line.

4. THE ESTIMATION OF THE SPECIFIC FREE SURFACE ENERGY $\gamma_{s,v}$ FOR POLYMERS

As shown in the Introduction, Good's theory uses the results of contact angle measurements in the relationship:

$$\Phi_{s,1}^2 \cdot \gamma_{s,v} = \frac{1}{4}(1 + \cos\theta_e)^2 \cdot \gamma_{1,v} \qquad (cf\ 9) \qquad (18)$$

Since the interaction parameter is not known, the surface energy $\gamma_{s,v}$ cannot be deduced directly. In general, a $\Phi < 1$ is expected [5]. Then, the maximum value of the term $\Phi^2 \cdot \gamma_{s,v}$ among a set of values obtained with several liquids would serve as a good approximation of $\gamma_{s,v}$. Table 1 presents such a set calculated from values of θ_e and $\gamma_{1,v}$ for PTFE as an example. Note that the set uses both polar and apolar liquids. The θ_e values were determined as described in Section 3, eqs. (14-17).

TABLE 1

Wetting liquid	θ_e [°]	$\Phi_{s,1}^2 \cdot \gamma_{s,v}$ [10^{-3} nm^{-1}]
n-decane	32.3	20.23
n-hexadecane	44.3	20.30 max.
α-bromo-naphthalene	69.0	20.29
methylene iodide	74.8	20.03
ethylene glycol	73.0	20.03
formamide	80.2	19.98
glycerol	83.0	19.95
water	94.4	15.51

For apolar liquids the values of $\Phi^2 \cdot \gamma_{s,v}$ are nearly identical and distinctly greater than those for polar liquids. Also PTFE consists of apolar molecules. This result confirms, qualitatively, the conclusion from Good's theory that phases with the same kind of molecular interaction should have a $\Phi \approx 1$. As expected we find $\Phi < 1$ for polar liquids on PTFE.

Further, Good's concept permits the testing of the quality of the θ_e values obtained with the method described in Section 3. From equation (3,9) it follows that:

$$\cos \theta_e = \frac{2 \cdot \Phi_{s,1} \sqrt{\gamma_{s,v}}}{\sqrt{\gamma_{1,v}}} - 1 \qquad (19)$$

As Table 1 reveals, it is possible to select a set of liquids so that $\Phi_{s,1}$ = constant with good accuracy. For $\gamma_{s,v}$ we take the maximum value of $\Phi^2 \cdot \gamma_{s,v}$. Then the plot of $\cos \theta_e$ against $(\gamma_{1,v})^{-\frac{1}{2}}$ should give a straight line cutting the ordinate at (-1). With the data of Table 1, we get (with the exception of water)

$$\cos \theta_e = \frac{9.296}{\sqrt{\gamma_{1,v} \quad 10^{-3} \frac{m}{N}}} - 1.044; \quad R = 0.9987$$

Following Good and other authors and setting $\theta_e = \theta_a$, we obtain

$$\cos \theta_a = \frac{12.257}{\sqrt{\gamma_{1,v} \cdot 10^3 \frac{m}{N}}} - 1.675; \quad R = 0.9803$$

In both cases, the linear fit works very well (compare the correlation coefficients R). This justifies the assumption of Φ = constant. However, our θ_e values meet the intersection of the ordinate at (-1) proposed by the theory much better. This result is a good argument for our way of finding θ_e. On the other hand this agreement favours Good's theory with respect to those of Zisman [15] and Lee [16] where the functions $\cos \theta_e = f (\gamma_{1,v})$ and $\cos \theta_e = f (\ln \gamma_{1,v})$ are expected to form straight lines.

Table 2 summarizes the specific free surface energies $\gamma_{s,v}$ of various polymers according to our determination.

TABLE 2

Polymer	PTFE	PE	EVA1-copolymer	PMMA	PS
$\gamma_{s,v}$ 10^{-3} $\frac{J}{m^2}$	20.3	27.5	33.8	39.4	33.2

REFERENCES

1. Adamson, W. and Ling, I., Adv. Chem. Ser., 1964, 43, 57.

2. Rusanov, A.I. "Phasengleichgewichte und Grenzflächenerscheinungen" Akademie-Verlag, Berlin 1978, p. 374.

3. Girifalco, L.A. and Good, R.J., "Theory for the Estimation of Surface and Interfacial Energies", J. Phys. Chem., 1957, 61, 904-909.

4. Good, R.J., "Surface Free Energy of Solids and Liquids", J. Coll. & Interf. Sci., 1977, 59, 398.419.

5. Good, R.J. in Lee, L.H. "Adhesion Science and Technology, Plenum Press, New York, 1975, vol. 9a, p. 107.

6. Johnson, R.E. and Dettre, R.H., "Contact Angle Hysteresis", Adv. Chem. Ser., 1964, 43, 112-144.

7. Busscher, H.J., Pelt. van, A.W.J. et al., "The Effect of Surface Roughening of Polymers on Measured Contact Angles of Liquids", J. Coll. & Surfaces, 1984, 9, 319-331.

8. Good, R.J. and Koo, M.N., "The Effect of Drop Size on Contact Angle", J. Coll. & Inferf. Sci., 1979, 71, 283.

9. Thiessen, P.A., and Schoon, E., "Benetzung und Adhäsionsarbeit fester organischer Verbindungen", Z.F. Elektrochemie, 1940, 46, 170-181.

10. Sakai, I., et al., "Degree of Interfacial Occupation of Polymer..", Kobunshi Ronbushu, 1978, 35, 209-214.

11. Wenzel, R.N., Ind. Eng. Chem., 1936, 28, 988.

12. Schulze, R.-D., Possart, W. and Kamusewitz, H., "Ermittlung des Gleichgewichtsrandwinkels an realen Festkörperoberflächen", Repr. 2nd Workshop on Adhesion, Institute of Polymer Chemistry, Academy Of Sciences of the G.D.R., 1985, p.205.

13. Bartell, F.E. and Shepard, J.W., "Surface Roughness as Related to Hysteresis of Contact Angles", J. Phys. Chem., 1953, 57, p. 211-215 and 455-458.

14. Kamusewitz, H. and Possart, W., "The Static Contact Angle Hysteresis Obtained by Different Experiments For the System PTFE/Water", Int. J. Adhesion & Adhesives, 1985, 5, 211.

15.   Zisman, W.A., "Relation of the Equilibrium Contact Angle to Liquid
      and Solid Constitution", <u>Adv. Chem. Ser.</u>, 1964, <u>43</u>, 1-51.

16.   Lee, M.C.H., "An Analytical Method for Determining the Surface Energy
      of Solid Polymers", General Motors Research Laboratories, GMR-4216,
      Michigan, 1983.

FACTORS AFFECTING THE DURABILITY OF TITANIUM/EPOXY BONDS

J. A. Filbey and J. P. Wightman
Department of Chemistry
Center for Adhesion Science
Virginia Polytechnic Institute
& State University
Blacksburg, VA   24061 USA

INTRODUCTION

Good durability of adhesive bonds in hot-wet environments is currently an area of active research [1-3]. Often, the weakest link in a hot, humid environment is the metal/adhesive or the metal/primer/adhesive interphases. These interphase regions need more study to determine the nature of the bonding between the adherend, primer, and adhesive.

The mechanisms by which materials adhere can be classified into four theories of adhesion: electrostatic, diffusion, mechanical interlock, and adsorption [1,4]. The electrostatic theory, primarily proposed by Deryaguin and coworkers [1], states that good adhesion is caused by the redistribution of charges that occurs when dissimilar materials come together. The diffusion theory involves polymer/polymer adhesion due to interdiffusion of the macromolecular chains at the interphase. The physical structure of the adherend causes good adhesion in the mechanical interlock theory. The adherend is structured so that adhesive can flow into crevices or ink-well shaped pores but not be easily pulled out. The fourth theory of adhesion, the adsorption theory, involves primary and secondary bonds formed at interphases to cause good adhesion. The secondary bonds are van der Waals forces which include London disperion forces, dipole-dipole, acid-base, and hydrogen bonding. The primary bonds are covalent or ionic bonds which form between the adherend and the primer or adhesive.

This paper reports the study of titanium oxide/epoxy and titanium oxide/metal alkoxide/epoxy interphases. The chemical and physical properties of the oxides were characterized. Next, the strength and durability of titanium/epoxy bonds was studied. Because some titanium

oxides were not durable in hot-wet environments, metal alkoxide primers
for adhesion promotion were investigated.

EXPERIMENTAL

Materials

Ti-6%Al-4%V was the only alloy used in this study. A 175° C cure
structural epoxy, FM-300U (American Cyanamid), was used for all adhesive
bonding. Chemicals, unless otherwise noted, were obtained from Fisher
Scientific Company.

Ti-6Al-4V Pretreatments

The Ti-6Al-4V adherends were pretreated by the following methods:
chromic acid anodization (CAA), sodium hydroxide anodization (SHA),
phosphate/fluoride acid etch (P/F), and TURCO basic etch (TURCO). The
procedures for these pretreatments are listed in the Appendix.

Spectroscopic and Microscopic Analysis

X-ray photoelectron spectra (XPS) were obtained on a Perkin-Elmer
PHI 5300 ESCA system using a Mg anode. Samples were punched as 0.95 cm
disks and scanned from 0 to 1200 ev. Narrow scans were made on any peaks
detected in the wide scan.

Auger electron spectroscopy (AES) was done on a Perkin-Elmer PHI 610
scanning Auger microprobe. An electron beam voltage of 3 to 5 kv and a
beam current of 0.05 μa was used. Samples were depth profiled by argon
ion sputtering with an ion beam voltage of 4 kv, an emission current of
25 ma, and an ion beam current of 0.2 μa.

Fourier transform infrared spectroscopy (FTIR) was used with a
grazing angle reflectance attachment to study alkoxide films on
reflective metal surfaces. Spectra were obtained on a Nicolet 5DXB
spectrometer using either a deuterated triglycine sulfate detector or a
high sensitivity liquid nitrogen cooled mercury cadmium telluride
detector. Peak positions were determined using cursor selected data
points on the computer.

Scanning transmission electron microscope (STEM) pictures were taken on a Philips EM-420T electron microscope. Ti-6-4 foil was pretreated and cut to 3 by 8 mm pieces. The samples were conductive, so coating the samples to prevent charging was not necessary.

Roughness

A Talysurf 4 profilometer determined roughness on a micrometer scale for the pretreated Ti-6-4 samples. A diamond stylus measured the average peak to valley profile. A glass slide served as the reference.

Surface Free Energy Determination

The polar, γ_s^P, and dispersive, γ_s^D, components of the surface free energy of the pretreated Ti-6-4 oxides were determined using an interfacial contact angle method used by Carre and Schultz on aluminum adherends [5]. The pretreated Ti-6-4 samples were immersed in a series of alkanes and the contact angle of a water drop measured on the immersed surface. The measured contact angles were corrected for roughness and porosity . A roughness coefficient, r, was obtained by first coating the pretreated surfaces and a smooth glass surface with gold. The contact angle of formamide was measured on both surfaces. By ratioing the cosine of the contact angle on titanium, θ_{Ti}, to the cosine of the contact angle on glass, θ_g, the roughness coefficient was obtained as shown in Equation (1).

$$\cos \theta_{Ti} / \cos \theta_g = r \tag{1}$$

The contact angles of water drops on pretreated Ti-6-4 obtained in the interfacial technique, θ_r, were then corrected for roughness by Equation (2) yielding a smooth contact angle, θ_s

$$[\cos \theta_r]/r = \cos \theta_s \tag{2}$$

The correction for a contact angle measured on a porous surface, θ_p, required a measure of the area of the surface covered with oxide and the area covered with pores, h. Equations (3) and (4) show the correction for the contact angles measured on a porous surface where N is the number of pores per square centimeter and R is the average radius of the pores.

$$\cos \theta_s = [((\cos \theta_p)/r) + h]/(1 - h) \tag{3}$$

$$h = N(\pi R^2) \tag{4}$$

Once the corrected contact angles were calculated, the dispersive component of the surface free energy was graphically determined using Equation (5).

$$[\gamma_w - \gamma_h + \gamma_{hw} \cos \theta_s] = 2(\gamma_s^D)^{1/2} [(\gamma_w^D)^{1/2} - (\gamma_h)^{1/2}] + I_{sw}^P \tag{5}$$

where γ_w is the surface energy of water, γ_h is the surface energy of alkane, γ_{hw} is the alkane/water interfacial energy, and γ_w^D is the dispersive component of the surface energy of water. The polar component of the Ti-6-4 surfaces was then calculated from the intercept, I_{sw}^P using both a geometric mean [6] and a harmonic mean [7] as shown in Equations (6) and (7), respectively.

$$I_{sw}^P = 2(\gamma_w^P \gamma_s^P)^{1/2} \tag{6}$$

$$I_{sw}^P = 4(\gamma_w^P \gamma_s^P)/(\gamma_w^P + \gamma_s^P) \tag{7}$$

Once the dispersive and polar components were obtained, the work of adhesion of an epoxy/Ti-6-4 bond in air and water were calculated by Equations (8) and (9), assuming values of 37.2 and 8.3 for γ_E^D and γ_E^P of the epoxy, respectively [8].

$$W_A = 2[(\gamma_{Ti}^D \gamma_E^D)^{1/2} + (\gamma_{Ti}^P \gamma_e^P)^{1/2}] \tag{8}$$

$$W_{AW} = 2[\gamma_w - (\gamma_{Ti}^D \gamma_w^D)^{1/2} - (\gamma_{Ti}^P \gamma_w^P)^{1/2} - (\gamma_E^D \gamma_w^D)^{1/2} -$$

$$(\gamma_E^P \gamma_w^P)^{1/2} + \gamma_{Ti}^D \gamma_E^D)^{1/2} + (\gamma_{Ti}^P \gamma_E^P)^{1/2}] \tag{9}$$

Alkoxide Coating Preparation

Three alkoxide coatings were used in this study - E-8385, sec-butyl aluminum alkoxide; TNBT, tetra n-butyl titanate; and TIPT, tetra isopropyl titanate, all received from Stauffer Chemical Company. One weight percent solutions in toluene were made in a glove bag purged with dry nitrogen. The toluene was dried by stirring over calcium hydroxide overnight. It was then distilled into a flask containing activated molecular sieves.

Five coats of solution were brushed on the adherends in a glove bag purged with dry nitrogen. For FTIR analysis, the alkoxide solutions were spun coat onto reflective metal substrates of either ferrotype plate or polished Ti-6-4. The alkoxide coating was cured at room temperature or at 300° C.

Adhesive Bonding

Lap shear coupons (2.54 x 12.7 x 0.13 cm) were pretreated and bonded with four layers of FM-300U epoxy. The pretreated surfaces were not primed with an organic primer. The joints were placed in a platen press and heated from room temperature to 175° C at 13.8 MPa bonding pressure for 1.5 hours. The heat was turned off and the joints allowed to cool to room temperature under pressure. All lap shear joints were aged for 10 days before testing. Lap shear bonds were pulled to break on an Instron at a crosshead rate of 0.127 cm per minute. Lap shear bonds were also loaded to 40% of breaking strength and placed in 80° C, 95% r.h. Time to failure was measured in a stress durability test.

Wedge coupons (2.54 x 15.24 x 0.38 cm) were pretreated and bonded with two layers of FM-300U using Teflon film spacers yielding a bond line thickness of 0.0381 cm. The pretreated surfaces were not primed with an organic primer. The joints were placed in a platen press and heated from room temperature to 175° C at 1.72 MPa bonding pressure for 1.5 hours. The heat was turned off and the joints allowed to cool to room temperature under pressure. A Ti-6-4 wedge was driven into one end of the wedge joint causing an initial crack to propagate. Samples were then placed into an environment of 80° C, 95% r.h. for 14 days. The CAA, SHA, and TURCO samples were also immersed in 95°C water. The position of the crack was measured with a ruler.

RESULTS AND DISCUSSION

Physical and Chemical Properties of Ti-6-4 Oxides

The four pretreated oxide surfaces differed in surface structure as seen by SEM and STEM [9]. The anodized surfaces were porous, with pore diameters of approximately 40 nm, while the P/F and TURCO surfaces showed no porosity. The roughness of these oxides on a micrometer scale also differed between pretreatments. Using profilometry, TURCO was the roughest surface with an average peak to valley height of 3.4 μm. The average peak to valley height for P/F was 2.8 and for CAA was 2.1. Considering both roughness and porosity, the anodized surfaces, CAA and SHA, had the highest surface area, followed by TURCO with a higher micrometer roughness than P/F.

XPS showed the chemical composition of the pretreated surfaces was reproducible. Carbon, oxygen and titanium were detected on all pretreated surfaces by XPS. CAA surfaces also contained a small percentage of fluorine, present as an inorganic fluoride from the electrolyte, HF. Chromium, in the anodizing bath, was not detected. The P/F surfaces contained a small percentage of phosphorus, from the etch solution. Fluorine was not usually detected. TURCO surfaces contained both silicon and iron. Venables and co-workers also detect iron particles on the TURCO pretreated surface [10]. When the surface was pickled after gritblasting but before the TURCO treatment, no silicon or iron was detected. XPS did not detect silicon or iron in the TURCO 5578 mixture. Iron and silicon, therefore, resulted from the gritblast step, and the pickling step removed the iron and silicon contamination. The SHA pretreatment for Ti-6-4 is reported by Kennedy, et al. [11]. Their optimum pretreatment conditions were chosen for this study. Silicon, calcium, phosphorous, and fluorine were detected on the SHA surface. The silicon was from the gritblasting step. The tap water rinse accounted for the calcium and the fluorine.

The polar and dispersive components of the surface free energy were determined for the CAA, P/F, and TURCO surfaces. Table 1 lists these values using both a geometric mean and a harmonic mean to calculate the polar component. From the polar and dispersive components of the surface energy, the work of adhesion of the epoxy/Ti-6-4 bond was calculated for air and water environments and is listed in Table 2. The calculated work of adhesion was significantly reduced for bonds in a water environment. Therefore, thermodynamics predicted a weaker bond between the titanium and epoxy in water than in air. The kinetics of moisture intrusion must also be considered.

Strength and Durability of Ti-6-4/FM-300 Bonds

The presence of a porous structure created by the anodization creates an excellent opportunity for increased surface area for adhesive/oxide interaction. It must be shown, however, that the epoxy penetrates the porous oxide.

FM-300 was cured on CAA pretreated Ti-6-4 foil. Because of the flexibility of the foil-epoxy sample, the two materials were peeled apart. Both surfaces were examined by STEM. While the metal failure side showed a faint evidence of porosity, it was unclear as to whether the surface was covered by a thin layer of adhesive or if the majority of the porous structure had been removed. The adhesive failure side appeared as if the epoxy had been pulled out of the porous structure. By XPS, titanium oxide was present on both surfaces with a high percentage of fluorine. The high fluorine concentration indicated fracture had occurred at the base of the porous structure [12]; however, epoxy penetration into the pores was still unclear. The adhesive failure was then depth profiled with AES. Figure 1 shows the atomic concentration with sputter time. The presence of the carbon signal simultaneously with the titanium and oxygen signals indicated that the epoxy did indeed penetrate the porous structure.

Strength to break measurements were made on lap shear joints made from CAA, P/F and TURCO pretreated Ti-6-4. The average breaking strengths for CAA, P/F and TURCO were 26, 25 and 28 MPa, respectively. With a standard deviation of ± 7 MPa, the breaking strengths were not significantly different, thus emphasizing the lap shear test of unexposed samples did not differentiate between surface pretreatments.

Whereas the lap shear strengths were not significantly different, the locus of failure differed depending on the pretreatment. For all pretreatments, the failure surfaces appeared to be a metal failure side (MFS) and an adhesive failure side (AFS). The failure surfaces were then studied by XPS. For the CAA pretreated coupons, both the MFS and the AFS showed titanium oxide present. A titanium metal peak, as well as an oxide peak, was also seen on the MFS. The oxygen peak occurred at 530.6 ev, indicative of a titanium oxide. The AFS contained a high percentage of fluorine. Thus, the bonds failed within the oxide layer, close to the base of the pores, a similar failure to the peel experiment.

The P/F and TURCO pretreated bonds showed a locus of failure different from the CAA bonds. Although the failure appeared to be interfacial, no titanium was detected on the AFS or the MFS. The oxygen peak at 533 ev indicated carbon-oxygen bonds in the epoxy. Bromine, present in the epoxy, was detected on both the AFS and the MFS. Thus, the P/F and TURCO bonds failed within the adhesive, close to the metal oxide.

Although the lap shear test was unable to differentiate between pretreatments, the stress durability test allowed the hydrolytic stability of the lap shear bonds to be tested while under load. A significant difference in time to failure was seen for the three pretreatments in the stress durability test as shown by the time to failure windows in Figure 2. The P/F surface showed poor durability with all bonds breaking in less than one day, while the TURCO and CAA surfaces showed better durability with the CAA being the most durable.

The wedge test was also sensitive to surface pretreatment. In the 80° C, 95% r.h. environment, the CAA, SHA, and TURCO samples showed no crack propagation in the test period. However, cracks propagated to failure in less than one day in the P/F samples. Figure 3 shows the crack propagation with time for the three pretreatments.

To accelerate the wedge test, samples were immersed in 95° C water and crack propagation was observed in samples which remained durable in the 80° C environments. After 16 days, the TURCO samples showed the initiation of crack growth, with the crack propagating 0.1 cm. The CAA and SHA samples showed no propagation at 16 days. However, for longer exposure times, cracks began to propagate in the CAA and SHA samples. Thus results from the stress durability test qualitatively agreed with results from the wedge test.

The locus of failure for the wedge and stress durability samples was determined using XPS. The MFS showed no presence of epoxy. Titanium, oxygen and carbon were detected . The oxygen binding energy was indicative of titanium oxide. The AFS showed only carbon, oxygen, nitrogen and bromine, with binding energies corresponding to that of the neat epoxy. The bonds failed at the epoxy/metal oxide interface. These results contrasted the locus of failure results obtained in unexposed lap shear samples.

Metal Alkoxides as Adhesion Promoters

Because of the poor durability of the P/F pretreatment, it was chosen as the substrate to test the efficacy of the alkoxide coating. Pike showed that sec-butyl aluminum alkoxide enhanced the durability of aluminum/epoxy bonds [13]. Figure 3 compares the durability of the alkoxide coated P/F samples to the uncoated P/F treated samples. The rate of crack propagation for the titanium alkoxide coated P/F samples was similar to that of the uncoated P/F samples. The two titanium alkoxides, TNBT and TIPT, showed no enhancement in bond durability of the wedge test over the P/F pretreatment. The E-8385 coated samples, on the other hand, showed a significant enhancement in wedge test bond durability of the P/F pretreatment. There was no crack propagation for two days, followed by slow crack growth.

Similar results were observed in the stress durability test for E-8385 on P/F. The loaded lap shear bonds failed between 10 and 20 days compared to less than one day for the uncoated P/F samples. The E-8385 coating appeared to mask the Ti-6-4 structure. STEM photomicrographs showed no difference between the P/F coated samples and the CAA coated samples, with the porosity of the CAA masked by the E-8385 coating.

The locus of failure of aluminum alkoxide primed P/F samples was investigated by XPS. Carbon, oxygen, titanium, aluminum and nitrogen were present on the MFS. The AFS contained carbon, oxygen, bromine and aluminum. Because aluminum was present on both failure surfaces, the failure occurred within the primer. The presence of nitrogen on the XPS presumably from the epoxy, indicated epoxy had interacted with the MFS.

Because of the durability of the titanium alkoxides differed dramatically from the aluminum alkoxides, the chemical differences between the alkoxide surfaces were investigated by FTIR. Figure 4 compares the OH stretching region for TNBT and E-8385. The concentration of hydroxyl groups on the E-8385 was significantly higher than on the TNBT surface.

SUMMARY AND CONCLUSIONS

The pretreatments given to Ti-6-4 created chemically and physically reproducible oxide surfaces. The anodized surfaces were porous, whereas, the etched surfaces were not. TURCO created a rougher surface than the P/F and CAA surfaces. The work of adhesion between epoxy and Ti-6-4 oxides is thermodynamically predicted to decrease in a water environment.

The strength and durability of Ti-6-4/FM-300 bonds was tested with the lap shear and the wedge tests. FM-300 penetrated the porous oxides. The lap shear test was not sensitive to the surface pretreatment when tested in ambient conditions. However, when the lap shear was placed under 40% load and placed in hot, humid conditions, a clear differentiation between surface pretreatments was observed. By both the stress durability and the wedge test, the CAA and SHA surfaces were most durable followed by the TURCO surfaces. The P/F surfaces showed the poorest durability. The locus of failure differed between the lap shear test and the durability tests.

Metal alkoxides were used to enhance the bond durability. Sec-butyl aluminum alkoxide significantly enhanced the bond durability of the P/F surfaces but leveled the durability of the CAA surface to that of the P/F coated with E-8385. The titanium alkoxides did not enhance the durability of P/F bonds. The locus of failure occurred within the primer layer. The aluminum alkoxide surface contained a higher concentration of hydroxyl groups than the titanium alkoxide surfaces. Therefore, the interphases with aluminum alkoxides have a higher density of hydrogen bonds than the titanium alkoxides.

One explanation for the differences in durability between the pretreatments is surface area. The porous oxides were the most durable and also contained the highest surface area. The TURCO surfaces were the roughest, therefore, increasing the surface area over that of the P/F surfaces. Moisture intrusion hydrolyzes physical bonds such as dipole-dipole, acid-base and hydrogen bonds. The kinetics of moisture intrusion is slower for the higher surface area surfaces, thus better durability. The metal alkoxide coated P/F samples, however, contain approximately the same surface area. But, the aluminum alkoxide is significantly more durable. The concentration of surface hydroxyls is higher, yielding a higher density of hydrogen bonds, thus better durability.

ACKNOWLEDGEMENTS

The authors would like to acknowledge the following people: Frank Cromer for training to operate the PHI XPS and AES systems and for help in data collection, Steve McCartney for operating the STEM.

The authors would also like to acknowledge the Office of Naval Research and NASA-Langley Research Center for financial support of the work. The gifts of Ti-6-4 from RMI Titanium, wedge coupons from the Martin Marietta Laboratory, FM-300U from American Cyanamid, the metal alkoxides from the Stauffer Chemical Company, and the stress durability tester from the 3M Company were appreciated. Enabling funds for the purchase of the surface analysis equipment were provided by the National Science Foundation and Virginia Tech.

REFERENCES

[1] A. J. Kinloch, ed. **Durability of Structural Adhesives**, Applied Science Publishers, London, 1983.

[2] W. Brockmann, O.-D. Hennemann, H. Kollek and C. Matz. <u>Int. J.</u>
     <u>Adhesion and Adhesives</u>, **6**, 115, 1986.

[3] M. Natan and J. D. Venables, <u>J. Adhesion</u>, **15**, 125, 1983.

[4] W. C. Wake, **Adhesion and the Formulation of Adhesives**, **2nd Edition**,
     Applied Science Publishers, London, 1982.

[5] A. Carre and J. Schultz, <u>J. Adhesion</u>, **15**, 151, 1983.

[6] D. K. Owens and R. C. Wendt, <u>J. Appl. Polymer Sci.</u>, **13**, 1741, 1969.

[7] R. J. Good, personal communication.

[8] J. Comyn, in **Durability of Structural Adhesives**, A. J. Kinloch,
     ed., Applied Science Publishers, London, 1983.

[9] J. A. Filbey and J. P. Wightman, submitted to <u>J. Adhesion</u>.

[10] B. M. Ditchek, K. R. Breen and J. D. Venables, Final Report, MML,
     TR-80-17-c, April, 1981.

[11] A. C. Kennedy, R. Kohler and P. Poole, <u>Int. J. Adhesion and</u>
     <u>Adhesives</u>, **3**, 133, 1983.

[12] M. Natan and J. D. Venables, MML, TR-82-20c, Sept. 1982.

[13] R. Pike, <u>Int. J. Adhesion and Adhesives</u>, **6**, 21, 1986.

APPENDIX

Chromic Acid Anodization (CAA)

1. Gritblast with an Econoline gritblaster at approximately 100 psi and held approximately 5 cm from the coupon.

2. Wipe with methyl ethyl ketone (MEK).

3. Soak in sodium hydroxide solution (13 g/250 ml) at 70° C for 5 minutes.

4. Rinse three times in deionized water.

5. Pickling step: Immerse in pickle solution (15 ml conc. HNO_3, 3 ml 49% w/w HF, 82 ml H_2O).

6. Rinse three times in deionized water.

7. Anodize at room temperature for 20 minutes at 10 volts at 27 A/m^2, in a chromic acid solution (50 g CrO_3/1000 ml) with Ti-6-4 as the cathode. 49% w/w HF is added to attain the desired current density.

8. Rinse three times in deionized water, soaking for 5 minutes in the final rinse.

9. Blow dry with prepurified N_2 gas until visibly dry.

Phosphate/Fluoride Acidic Etch (P/F)

1. Gritblast as in CAA.
2. Wipe with MEK.
3. Soak in Sprex AN-9 solution (30 g/1000 ml) at 80° C for 15 minutes.
4. Rinse three times in deionized water.
5. Immerse in pickle solution (31 ml 49% w/w HF, 213 ml conc. HNO_3/1000 ml) at room temperature for 2 minutes.
6. Rinse three times in deionized water.
7. Soak in phosphate/fluoride solution (50.5 g $NaPO_4$, 20.5 g KF, 29.1 ml 49% w/w HF/1000 ml) at room temperature for 2 minutes.
8. Rinse three times in deionized water.
9. Soak in deionized water at 65°C for 15 minutes.
10. Blow dry as in CAA.

TURCO Basic Etch

1. Gritblast as in CAA.
2. Wipe with MEK.
3. Soak in TURCO 5578 solution (37.6 g/1000 ml) at 70-80° C for 5 minutes.
4. Rinse three times in deionized water.
5. Soak in TURCO 5578 solution (380 g/100 ml) at 80-100° C for 10 minutes.
6. Rinse three times in deionized water.
7. Soak in deionized water at 60-70° C for 2 minutes.
8. Blow dry as in CAA.

Sodium Hydroxide Anodization (SHA)

1. Rinse with methanol or acetone.
2. Immerse in Super Terj (30 g/1000 ml) at 80° C for 15 minutes.
3. Soak in water at 50-60° C for 15 minutes.
4. Anodize at 20° C for 30 minutes at 10 volts in 5.0 M sodium hydroxide solution with Ti-6-4 or stainless steel mesh as the cathode. The current density was not controlled.
5. Rinse in running tap water for 20 minutes.
6. Dry in oven at 60° C for 10 minutes.

TABLE 1

Dispersive and polar components of the surface energy (in mJ/m^2) of pretreated Ti-6-4, using the geometric mean (GM) and the harmonic mean (HM).

Pretreatment	γ^D	γ_s^P (GM)	γ_s^P (HM)
CAA	87	30	32
P/F	64	14	18
TURCO	52	16	20

TABLE 2

Work of adhesion (in mJ/m^2) of Ti-6-4/epoxy bonds in air (W_A) and in water (W_{AW}).

Pretreatment	Mean	W_A	W_{AW}
CAA	GM	146	26.6
CAA	HM	140	33.8
P/F	GM	119	37.7
P/F	HM	120	51.1
TURCO	GM	112	32.8
TURCO	HM	112	45.2

31

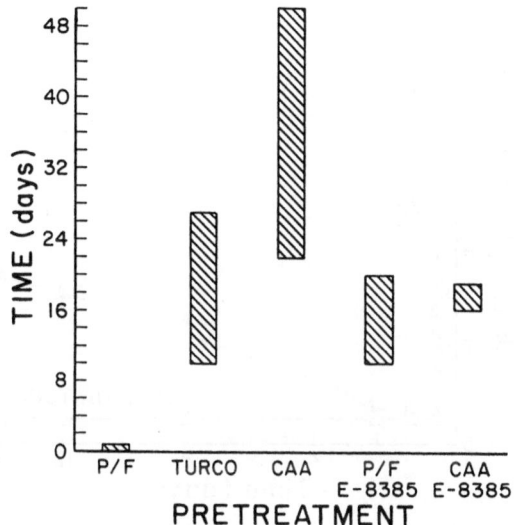

FIGURE 1. Auger depth profile of adhesive failure surface.

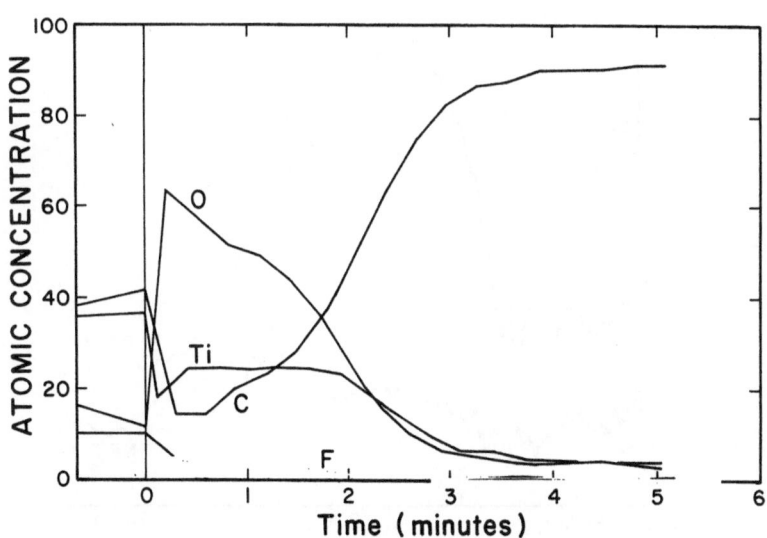

FIGURE 2. Time to failure windows for the stress durability test at 40%

strength to break load, 80° C, 95% r.h.

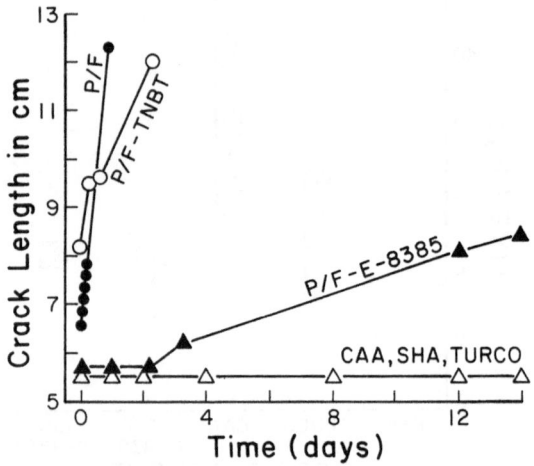

FIGURE 3. Crack length (in cm) vs. time (in days) of wedge samples at 80° C, 95% r.h.

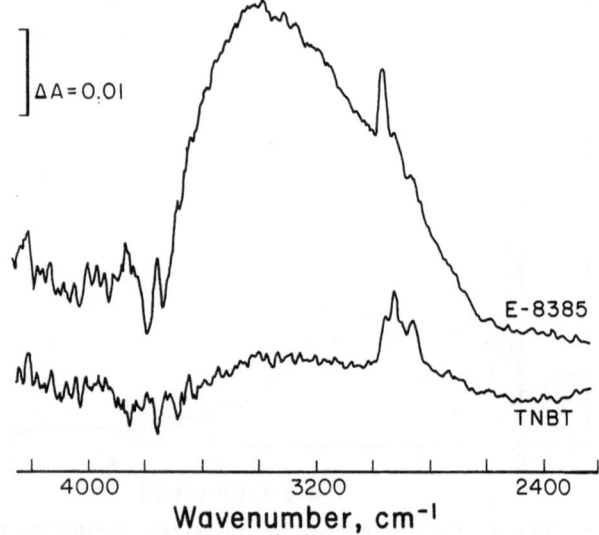

FIGURE 4. FTIR reflectance spectra of -OH stretch region.

THE RÔLE OF ADHESION IN MINERAL FILLED PLASTICS

P.J. Wright
Blue Circle Industries PLC,
Greenhithe, Kent,
UK

INTRODUCTION

The physical properties of plastics can be considerably changed by the incorporation of mineral fillers. The addition of any mineral filler will increase the modulus of the plastic, reduce the dependance of the mechanical properties on temperature and reduce shrinkage. Fibrous and plate-like fillers can also significantly increase tensile and flexural strength as well as giving a greater degree of improvement in the other mechanical properties[1].

The transfer of stress between filler and matrix requires adequate interfacial bonding. Interfacial friction is not sufficient to achieve full transfer and organo functional silanes[2], titanates[3] or other coupling agents[4,5] are used to modify the filler surface. These materials increase matrix filler interactions by both physical and chemical means.

The nature of the interface between filler and polymer plays an important part in determining many aspects of composite performance. In the first stages of composite preparation the surface treatment may aid dispersion of the filler in the plastic and protect it from damage during mixing.

The presence of polymer-filler interactions is essential in preserving the mechanical properties of composites in adverse environmental conditions. Chemical bonding acts to prevent the full interaction of a liquid phase with both components of the composite so that the work of adhesion at the interface remains the same in the presence or absence of the liquid[6].

The rôle of adhesion in altering the impact strength of a composite is complicated. The results obtained from the impact testing of plastics depend on specimen dimensions, the geometry of the notch, if present, and the strain rate of the test[7]. The material variables which affect impact strength include filler volume fraction, particle geometry, filler modulus and strength and matrix toughness.

For brittle plastics containing spherical or irregularly shaped fillers measured changes in fracture energy related to adhesion are not reflected in impact strength. There is, however, a direct correlation between the toughness of the composite taken as area under flexural stress-strain curves and impact strength[8]. The addition of a spherical filler to a tough thermoplastic degrades the energy absorbing capacity of the material and reduces both fracture energy and impact strength[9].

In short fibre filled plastics energy can be absorbed as fibres are debonded and pulled from the matrix as the composite breaks into two or more pieces[10,11]. In addition to all the other variables controlling impact strength fibre alignment will be important and there may be an optimum level of adhesion beyond which the crack will propagate through both fibre and matrix resulting in no increase in fracture energy.

To relate the observed performance of a composite directly to filler/matrix adhesion, full account must be taken of parameters such as volume fraction filler, the orientation of the particles and their geometry and the properties of the matrix. Most of the published work, particularly that which describes the relationship between reinforcing fillers and composite strength has been based on the use of short glass fibres. In this investigation the influence of adhesion on the strength, stiffness and impact behaviour of thermoplastic composites has been examined using acicular (needlelike) wollastonite, a naturally occurring calcium metasilicate.

Wollastonite, although it does not have the same potential for reinforcement as glass fibre can be used to provide composites having a useful balance of properties particularly in its surface coated forms[12].

THEORY

List of Symbols

E	Youngs modulus
σ	tensile stress
$\overline{\sigma}_f$	average fibre stress
σ'_m	tensile stress in matrix at fibre failure strain
τ_{fm}	fibre/plastic interfacial shear strength
η	orientation factor
l	actual fibre length
l_c	critical fibre length
d	actual fibre diameter
A	actual aspect ratio
A_c	critical aspect ratio
V_f	volume fraction of fibre
V_m	maximum packing fraction of filler

Subscripts   f filler   m matrix   c composite

Composite Strength

In the case of a continuously reinforced fibre composite in which the fibres are completely aligned in the direction of the applied load the two components will deform equally provided there is good adhesion between fibre and matrix. The tensile strength of such a composite is given by a simple law of mixtures model

$$\sigma_c = \sigma_f V_f + \sigma'_m (1 - V_f) \tag{1}$$

Kelly[13,14] modified equation (1) to account for short fibre composites which do not have uniform stress along the fibres by introducing the concept of critical fibre length. If a fibre can be withdrawn from the matrix by a tensile load only slightly smaller than the fibre breaking load the embedded length must be equal to half the critical length, $l_c/2$, since only one fibre end is concerned. The critical aspect ratio l_c/d can then be written as

$$\frac{l_c}{d} = A_c = \frac{\sigma_f}{2\tau_{fm}} \tag{2}$$

This relationship demonstrates the importance of bonding between fibre and matrix because the lower the interfacial shear strength (τfm) the longer the fibre, or the higher the aspect ratio, required to transfer stress. If a linear variation of tensile stress in the fibre with distance from the fibre ends can be assumed the mean fibre stress at failure is

$$\overline{\sigma}_f = \left(1 - \frac{\sigma_f}{4 A \tau_{fm}} \right) \sigma_f \qquad (3)$$

The composite tensile strength is therefore

$$\sigma_c = \sigma_f V_f \left(1 - \frac{\sigma_f}{4 A \tau_{fm}} \right) + \sigma_m \left(1 - V_f \right) \qquad (4)$$

$$A > A_c$$

If the fibres have an aspect ratio lower than the critical value the maximum fibre stress is, from equation (2), $2\tau f_m A_c$ and since the mean stress is half this value the composite tensile strength is

$$\sigma_c = \tau_{fm} A V_f + \sigma_m' \left(1 - V_f \right) \qquad (5)$$

$$A_c < A$$

If the actual aspect ratio is about ten times greater than the critical aspect ratio equation (4) becomes identical with equation (1). The low values of composite tensile strength predicted by equation (5) show the importance of a high aspect ratio if the composite is to make full use of the tensile strength of the fibre.

The variations in composite strength with filler volume fraction predicted by equations (4) and (5) are rarely obtained in practice. In the normal process of injection moulding or extrusion the alignment of the fibres is poor. To enable equations to be applied in these practical situations full allowance must be made for fibre orientation both in the plane and through the thickness of the moulding[15,16].

In such mouldings there is also a spectrum of aspect ratios. The contributions to composite strength from fibres of sub-critical and super-critical aspect ratio must be separately summed over the effective range of aspect ratios. A model has been suggested by Bader and Bowyer[17] which allows for a distribution of aspect ratios and for fibre orientation.

$$\sigma_c = \eta \left\{ \sum_i \tau_{fm} A_i V_i + \sum_j \sigma_f V_i \left(1 - \frac{\sigma_f}{4 A_j \tau_{fm}} \right) \right\} + \sigma'_m (1 - V_f) \quad (6)$$

If there is no adhesion between the matrix and a uniformly dispersed spherical filler the tensile yield stress of the composite is given by a simple model which assumes that all of the stress is borne by the matrix[18].

$$\sigma_c = \sigma_m \left(1 - V_f^{2/3} \right) \quad (7)$$

Where f will depend on the maximum packing that can be achieved by the filler.

Composite Modulus

There are many theoretical and empirical models which can be used to describe the moduli of composites[19]. In most of these models perfect adhesion is assumed between filler and polymer matrix, which is only true at very low strains for composites containing uncoupled fillers.

The simplest case assumes equal strains in the two phases under elastic deformation and is a simple law of mixtures model giving the upper bound for composite modulus as

$$E_c = E_f V_f + E_m (1 - V_f) \quad (8)$$

Equations of this type are only applicable to small concentrations of spherical fillers. A few of the models attempt to predict the effect of particle shape and concentration over a wide range. The model proposed by Nielsen[20,21,22] has been accepted as being one of the most accurate over a wide range of filler parameters.

$$E_{11} = E_m \left[\frac{1 + 2A B V_f}{1 - B \psi V_f} \right] \quad (9)$$

where $\psi = 1 + \frac{(1 - V_m)}{V_m^2} \cdot V_f$ $\qquad B = \frac{[E_f / E_m] - 1}{[E_f / E_m] + 2A}$

This model can also be used to describe the thermal conductivity and the viscosity of composites.

Models of this type can be modified to take account of aspect ratio distribution and used to calculate composite moduli in a manner which takes full account of the fibre orientation distribution found in practice[23].

If both composite modulus and fibre orientation can be accurately measured and fibre modulus and matrix modulus are known, some of the models can also be used to calculate a number average mean for the aspect ratio of the fibres[24].

Recently a number of models have been developed in which the physicochemical behaviour of a filler-matrix interlayer or mesophase is of importance[25,26,27,28]. It is necessary for the development of these models to consider adhesion at the interface between filler and mesophase and mesophase and matrix. The relationship between adhesion and mesophase thickness is such that when there is perfect adhesion and an equality of strains in the two phases the thickness of the interphase reduces to zero.

The existence of a bound polymer layer which has different properties from the bulk polymer has been clearly demonstrated[29] but the volume fraction of this layer will be low except for a few mineral fillers of high surface area used at high filler loadings.

The effect of the bound polymer layer on modulus, strength or the impact performance of most composites will therefore be minimal.

MATERIALS

Wollastonite Characterisation

The filler used for this work was an acicular wollastonite of Indian origin. The tensile strength and modulus of fibres were determined by AERE, Harwell, using a micro tensile testing machine[30]. This requires specimens of at least 2.5mm length. The usual range of mean fibre diameter of wollastonite grades used in reinforcing application is from 500 μm to less than 5 μm. To obtain fibres with diameters less than 65 um yet with lengths of 2.5mm presented a difficult problem as the frequency of such fibres is an insignificant part of the fibre population. The method of electrohydraulic crushing was used in which shock waves are introduced into a suspension of particles by a series of controlled electrical discharges. Using this technique fibres were generated with diameters down to 15 μm yet with length in excess of 2.5mm (as aspect ratio of over 167:1).

Tensile testing was carried out on these fibres and tensile strength and modulus measured for each. The mean of the tensile moduli was found to be 117 (± 15) GPa. The tensile strength values were plotted against equivalent fibre diameter. The straight line relationship between tensile strength and fibre diameter is similar to other synthetic fibre systems. From this relationship it appears that tensile strenths between 800 MPa and 1700 MPa might be expected for wollastonite fibres with diameters between 15 μm and 5 μm. The grade of wollastonite used for this work had a mean fibre length of 90 μm and an aspect ratio of between 10:1 and 20:1.

Filler Coatings

The coupling agent used to give adhesion between the filler and nylon 6 was δ - aminopropyltriethoxy silane (Union Carbide A1100). A mixture of organosilicon chemicals (PC1A/2B) recently developed by Union Carbide was used to give adhesion between wollastonite and the polyolefins[31]. This coupling system avoids some of the problems associated with the azidosilanes, the only coupling agents for polyolefins previously available[32].

Wollastonite was also coated with stearic acid to aid its incorporation and dispersion into the polyolefins. This treatment does not improve adhesion to the matrix.

Filler Coating Methods

Wollastonite was coated with 0.5% w/w of the amino silane by treating the filler in a high speed mixer with the silane diluted 1:1 using a mixture of methanol and water (9:1 v/v). The coated filler was dried at 105°C. The coupling agents for the wollastonite-polyolefin composites were not used to pretreat the filler but were added to both the filler and the polymer while mixing in a ribbon blender. 2.0% w/w of PC1A (based on filler weight) and 0.5% w/w of PC1B were used for the polypropylene system and 2.0% w/w of PC1A and 1.0% w/w of PC1B for the high density polyethylene system.

SPECIMEN PREPARATION

Extrusion Compounding

Dry blends of the fillers and the polyolefins were prepared using a ribbon blender and then fed to a Werner and Pfleiderer ZSK 30 twin screw compounding machine. For nylon the blends were fed to a Baker Perkins MPC V/30 twin screw compounding machine. In each case the extrudate in the form of strands, was cooled by passing through a water bath and then granulated. The unfilled polymers were also passed through the compounding machines.

Injection Moulding

Test specimens were moulded using a 35 tonne Szekekely in-line reciprocating screw injection moulding machine. Three types of specimen were moulded, tensile specimens according to ISO R527, Charpy bars according to ASTM D256 and plaques 100 x 100 x 2.5mm. These were used both for falling weight impact tests and for the preparation of small tensile test specimens used to examine fibre orientation effects.

The filler contents of some of the specimens were measured by ignition of the polymer at 500°C and the recovered fibre was used to determine the loss of aspect ratio occurring in the compounding and moulding process.

SPECIMEN TESTING

The testing of the injection moulded specimens was carried out according to the test procedures given in Tables 2, 3 and 4. To obtain accurate measurements of composite modulus which take full account of fibre orientation small dumb-bell specimens were cut at 0° and 90° to the flow direction from 2.5mm thick edge-gated plaques. Three specimens were cut from each plaque and five plaques were used to provide the specimens for each flow direction.

A plot of secant modulus as a function of strain was made using tensile stress-strain measurements. A typical example, for a nylon 6 - wollastonite composite containing 50% w/w of the filler is shown in Figure 1. The values of composite modulus at low strain obtained from these measurements were used to calculate the mean aspect ratio of the fibre in the composite taking full account of the fibre orientation throughout the thickness of the moulding. The fibre orientation found in polypropylene composites and its dependance on moulding thickness is shown in Figure 2.

Unnotched impact tests were performed using an instrumented falling weight machine. The drop height was 1.3m producing an impact velocity of 5 m/s. The total mass of the impactor was 15 Kg and the impact energy 191J. The impactor was hemispherical having a diameter of 16mm. A full force deflection curve was obtained from the machine enabling both energy to peak force and total failure energy to be calculated. the total failure energy is given in the tables. All the specimens were tested in the dry, as moulded, condition.

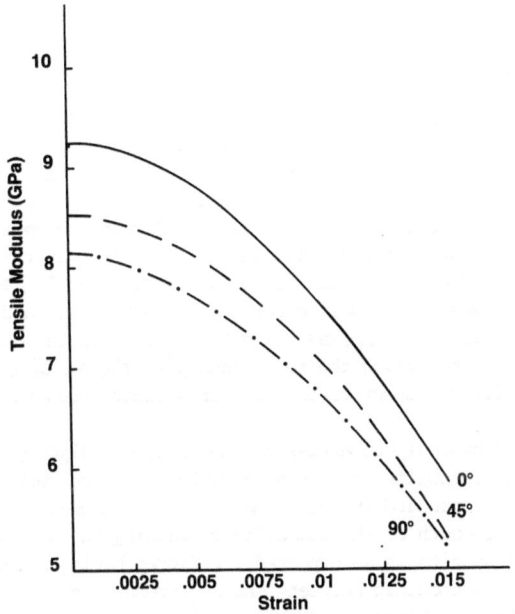

Figure 1. Secant modulus plotted as a function of strain using specimens
cut at 0°, 90° and 45° to the direction of mould fill from
2.5mm nylon 6 plaques containing amino silane coated
wollastonite

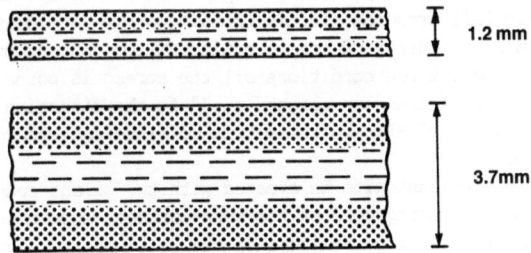

Figure 2. Schematic representation of fibre orientation in mouldings.
Viewed in the injection direction

RESULTS

Polypropylene-Wollastonite Composites

The polypropylene used for this work was a homopolymer having a
glass transition temperature, Tg, of 12°C. Polypropylene of this type
will fail in a brittle mode in most impact tests.   The low stiffness,
strength and heat deflection temperature of polypropylene can only be
improved by the incorporation of fillers and reinforcements. The mean
aspect ratio of wollastonite will be considerably less than the critical
value for the aspect ratio calculated using Equation 2. The fibre tensile strength was measured as described in section 3.1 and the interfacial shear strength, τ_{fm} was taken to be equal to the matrix shear
strength calculated using the relationship[33] $\tau_m = \sigma_m \sqrt{3}$. The
values of critical aspect ratio and matrix shear strength are given in
Table 1.

The mechanical properties of the unfilled polypropylene and the
polypropylene composite are given in Table 2. These results have been
obtained using standard injection moulded test pieces. The small
increase in strength of the composite containing the silane treated
coupled wollastonite is as expected for fibres having a low aspect
ratio. The relationship between tensile strength and volume fraction
filler is shown in Figure 3. These results have been obtained using the
small tensile test pieces cut at 0° and 90° to the direction of mould
fill from 2.5mm thick plaques. Polypropylene - wollastonite composites
containing 20, 30, 40 and 50% w/w of the stearate coated filler and a
composite containing 40% w/w of the silane coated filler were used in
these tests.

The results obtained from the specimens cut at 90° to the direction of mould fill from the plaques containing stearate coated wollastonite agree with the decay of strength with filler content predicted by
Equation 7. Under these conditions all the stress is borne by the
matrix. When the fibres are aligned at 0° to the direction of mould
fill some transfer of stress from matrix to fibre can occur by frictional contact.

When the wollastonite is treated with the silane coupling agent
scanning electron micrographs of fracture surfaces show that failure
occurs in the matrix or by fracture of the wollastonite along cleavage
planes (Figure 4), This indicates that the interfacial shear strength
is at least equal to matrix shear strength. The tensile strengths of
these composites agree well with the predictions of Equation 5 if a value of 6.5:1 is used for the aspect ratio of the wollastonite and
allowance is made for the existance of the skin-core structure of the
type shown in Figure 2.

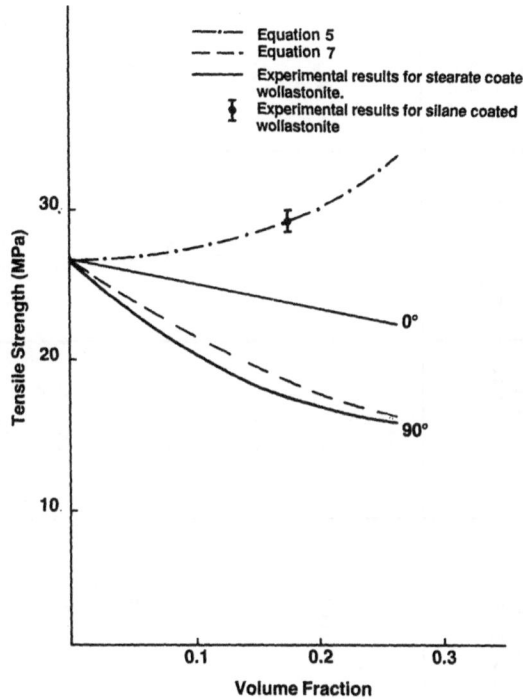

Figure 3. Tensile strength plotted as a function of volume fraction using specimens cut from 2.5mm polypropylene plaques

TABLE 1
Critical Aspect Ratios

Matrix Material	Matrix Shear Strength	Critical Aspect Ratio
Nylon 6	49	10:1
Polypropylene	18	28:1
High Density Polyethylene	13	37:1

TABLE 2

Mechanical Properties of Polypropylene-Wollastonite Composites

Property	Test Method	Units	Unfilled Polypropylene	Acicular Wollastonite	Acicular Wollastonite Silane Coated	Acicular Wollastonite Stearate Coated
Tensile Strength at yield	ISO R527	MPa	31	28	35	28
Tensile Elongation at break	ISO R527	%	>600	22	20	400
Tensile Modulus	ISO R527	GPa	1.1	2.2	2.0	2.5
Flexural Strength	ISO R178	MPa	40	44	58	42
Flexural Modulus	ISO R178	GPa	1.8	3.9	3.8	3.9
Heat Deflection Temperature	ISO R75 (method A)	°C	62	80	85	80
Notched Izod Impact Strength	ASTM D256A	$J.m^{-1}$	20	15	18	19
Unnotched Impact Strength	*	N.m	2.0	1.1	2.0	3.9

* See section on Specimen Testing.

The measurements of secant modulus as a function of strain made using the specimens cut from the plaques confirm that modulus at low strains, <0.2% does not depend on the existance of bonding between fibre and matrix. The behaviour of stearate coated wollastonite in greatly increasing the maximum tensile elongation at break is similar to the manner in which stearate coated calcium carbonate causes a shift from brittle to ductile failure in polypropylene composites, with most of the energy being absorbed by plastic deformation before failure occurs[34].

Heat distortion temperature is a measure of flexural stiffness made at strains which are sufficiently high to show small effects due to coupling between matrix and filler. In this case bonding the wollastonite to the polypropylene matrix increases the heat deflection temperature by 5°C. Measurement of modulus over the full temperature range of interest using dynamic mechanical methods can provide much more detailed information than these possibly misleading single point measurements[35].

The low notched impact strength of the unfilled polypropylene is not improved by the presence of wollastonite fibres as they are not sufficiently long to enable energy to be expended in pull-out and debonding processes. The unnotched impact strengths illustrate the value of coupling between filler and matrix. The already low impact strength of homopolymer polypropylene is further reduced by the presence of the uncoated wollastonite, each fibre acting as a discontinuity capable of initiating cracks. Scanning electron micrographs of the fracture surface of composites containing uncoupled, acicular wollastonite such as Figure 5 show that when there is no bonding between fibre and matrix the crack propagates from fibre to fibre. When there is good interfacial bonding the cracks propagate in a fibre avoidance mode through the matrix. The increased impact strength of the composite containing stearate coated wollastonite is explained by the same plastic deformation mechanism which cause the increased elongation at break found in the tensile test.

Nylon 6 - Wollastonite Composites

The polyamides are one of the most important groups of plastics used in engineering applications. Reinforcements such as glass fibre, glass beads, talc and mica are used to increase stiffness and heat distortion temperature. The particulate mineral fillers are often used when the pronounced anisotropic effects, present when the longer fibres are used, are unacceptable. Both nylon 6 and nylon 6.6 absorb a large amount of water at equilibrium. The use of a low water absorption

46

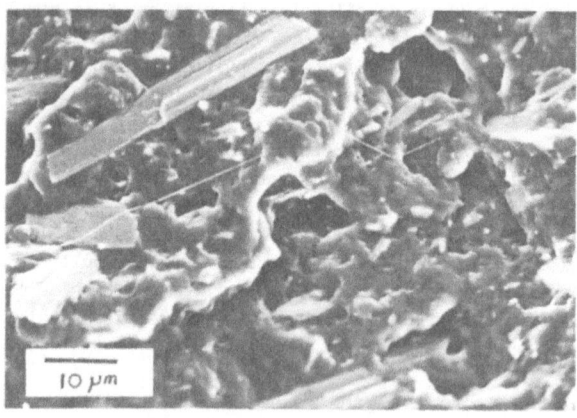

Figure 4. Fracture surface of an unnotched impact test specimen containing acicular wollastonite. The wollastonite and polypropylene blend having been treated with silane PC1A/PC1B

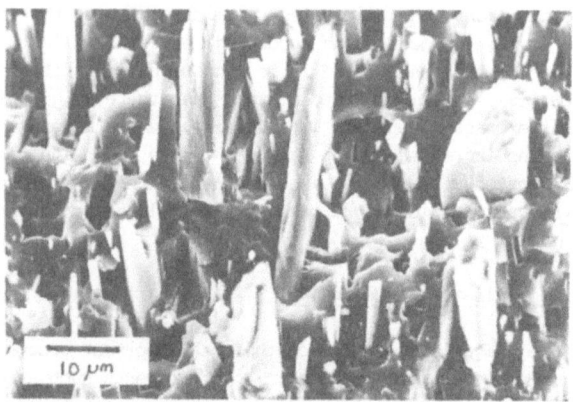

Figure 5. Fracture surface of an unnotched impact test specimen containing uncoated acicular wollastonite

filler such as wollastonite helps to preserve the properties of the polyamide in high humidity environments, but the presence of stable interfacial bonds between filler and matrix is essential.

Because of the high shear strength of the nylon 6 matrix the calculated value of the critical aspect ratio for wollastonite is close to the measured aspect ratio of the filler present in the composite. The measured tensile strengths given in Table 3 confirm both the value of achieving a bond strength between matrix and filler equivalent to the matrix shear strength and of using a filler having a significant aspect ratio to enable stress to be transferred from the matrix. The use of the uncoated particulate wollastonite having an aspect ratio of approximately 1:1 results in all the stress being borne by the matrix. Treatment of this filler with an amino silane coupling agent increases the composite strength to the level given by the uncoated acicular wollastonite where stress transfer occurs by frictional forces.

Figure 6 shows the relationship between composite tensile strength and volume fraction loading of the acicular wollastonite. The values of tensile strength at yield for the composite containing the coupled filler, measured at 0°, agree well with the predicted values calculated using Equation5. The calculations were made using a value for the aspect ratio of 6.5:1, obtained from modulus measurement, and by assuming that the matrix-filler bond strength was equivalent to the shear strength of the matrix. Scanning electron micrographs of the fracture surfaces of the tensile test pieces of which Figure 7 is a typical example show that failure has occurred in the matrix and that all the fibres pulled from the matrix are coated with nylon. The fracture surfaces of composites containing the uncoated wollastonite have a similar appearance to the fracture surface of the polypropylene composite shown in Figure 5.

The modulus measurements made as part of the stress-strain tests performed on the specimens cut from the plaques again show that when modulus is determined at strains less than 0.2% there is no difference between composites containing equal volume fractions of the uncoated wollastonite and the silane coated wollastonite. The small differences which do exist at strains of 0.2% to 0.3 are shown in the small increase in heat deflection temperature found when comparing the composites containing the coupled and uncoated wollastonite.

The notched impact strength of the composite containing the silane coated acicular wollastonite is significantly improved when compared to the composite containing the silane coated non-acicular wollastonite but the fibres are not of sufficiently high aspect ratio

TABLE 3

Mechanical Properties of Nylon 6 – Wollastonite Composites

Property	Test Method	Units	Unfilled Nylon 6	Non Acicular Wollastonite	Non Acicular Wollastonite Silane Coated	Acicular Wollastonite	Acicular Wollastonite Silane Coated
Tensile Strength at yield	ISO R527	MPa	81	44	66	65	96
Tensile Elongation at break	ISO R527	%	60	2.4	3.3	1.3	3.0
Tensile Modulus	ISO R527	GPa	2.7	4.1	4.0	5.6	6.0
Flexural Strength	ISO R178	MPa	108	86	108	118	166
Flexural Modulus	ISO R178	GPa	3.7	4.7	4.6	9.8	10.8
Heat Deflection Temperature	ISO R75 (method A)	°C	60	105	114	185	193
Notched Izod Impact Strength	ASTM D256A	$J.m^{-1}$	29	16	18	22	30
Unnotched Impact Strength	*	N.m	No break	32	60	29	88

* See section on Specimen Testing

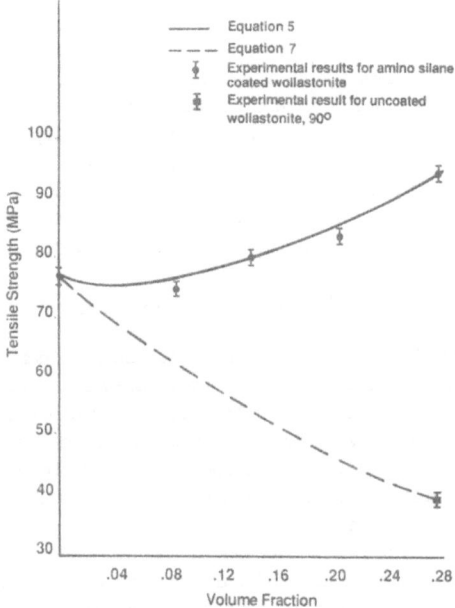

Figure 6. Tensile strength plotted as a function of volume fraction
using specimens cut from 2.5mm nylon 6 plaques

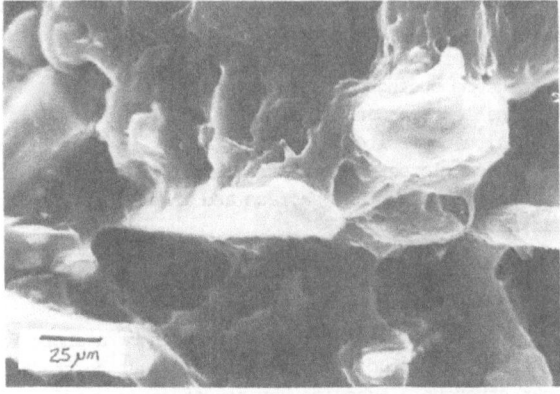

Figure 7. Fracture surface of a tensile test specimen containing
silane coated acicular wollastonite

and are too brittle to increase the impact strength of the composite
above that of the unfilled polymer. The unnotched impact strength of
nylon 6 is high and is drastically reduced by the removal of tough
polymer and its replacement with a particulate mineral filler. The presence of a short fibre which is not bonded to the matrix has a particularly deleterious effect on unnotched impact strength. Impact
measurements made using a wide range of volume fraction filler show that
at low levels the voids associated with the fibres act as sharp notches
with cracks running from fibre to fibre creating a fracture surface
covered with exposed fibres and the holes from which the fibres have
been removed.

High Density Polyethylene - Wollastonite Composites

The results obtained for the HDPE-wollastonite composites are
similar to those obtained for the polypropylene-wollastonite composites.
An interesting feature of these results given in Table 4 is the
increased stiffness and heat distortion temperature of the composites
containing the stearate coated wollastonite. The differences in modulus
are shown more clearly in the stress-strain measurements made using the
small tensile specimens cut from plaques. Figure 8 shows that at low
strains there is no difference between the modulus of composites containing uncoated and silane treated wollastonite but that the modulus of
the composite containing stearate coated wollastonite is considerably
higher. The mean aspect ratio of the wollastonite present in the specimens was determined after ashing the composite. For both the uncoated
and silane treated wollastonite the mean fibre length and the mean
aspect ratio of 6:1 were the same. The stearate coated wollastonite had
a longer fibre length and an aspect ratio of 8:1. It seems that the
stearate coating protects the fibre from breakage during the initial
stages of the compounding process.

The excellent bonding of matrix to fibre in the silane treated
HDPE-wollastonite composite is shown in Figure 9. This is a typical
scanning electron micrograph of a fractured unnotched impact specimen.
Few fibres are visable on the surface and those that are, are covered
with matrix material.

CONCLUSIONS

To achieve the maximum tensile strength of a short fibre filled
thermoplastic the interfacial shear strength must be greater than the
matrix shear strength. Adhesion has no effect on stiffness at very low
strains, the effect of adhesion only starts to be apparent at strains of
above 0.2%. Properties related to stiffness such as heat deflection
temperature will also be only slightly affected by the level of adhesion.

TABLE 4
Mechanical Properties of HDPE - Wollastonite Composites

Property	Test Method	Units	Unfilled HDPE	Acicular Wollastonite	Acicular Wollastonite Silane Coated	Acicular Wollastonite Stearate Coated
Tensile Strength at yield	ISO R527	MPa	23	22	32	26
Tensile Elongation at break	ISO R527	%	>600	32	13	10
Tensile Modulus	ISO R527	GPa	1.7	5.0	4.5	6.5
Flexural Strength	ISO R178	MPa	26	34	42	42
Flexural Modulus	ISO R178	GPa	1.1	3.1	3.1	4.9
Heat Deflection Temperature	ISO R75 (method A)	°C	45	64	63	75
Notched Izod Impact Strength	ASTM D256A	$J.m^{-1}$	25	18	20	17
Unnotched Impact Strength	*	N.m	3.2	3.5	4.5	3.4

* See section on Specimen Testing

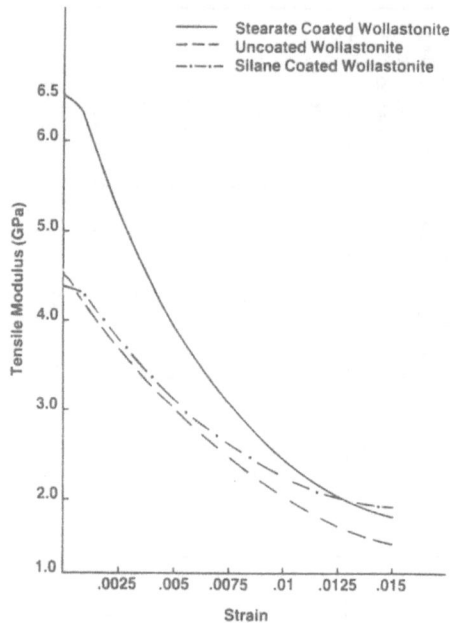

Figure 8. Secant modulus plotted as a function of strain using specimens cut at 0° to direction of mould fill from high density polyethylene plaques containing stearate coated wollastonite ($V_f = 0.17$)

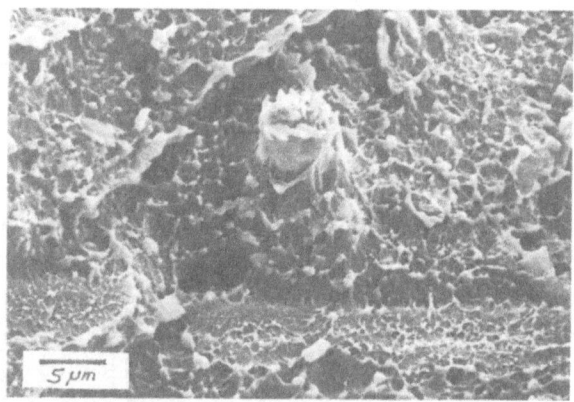

Figure 9. Fracture surface of a tensile test specimen containing acicular wollastonite. The wollastonite and HDPE blend having been treated with silanes PClA/lB

The unnotched impact strength of composites containing low aspect ratio fillers can be considerably improved by a high level of adhesion between matrix and filler. The notched impact strengths are not significantly affected by adhesion when the aspect ratio of the filler is below its critical value.

REFERENCES

1. Katz and Milewski, Handbook of Fillers and Reinforcements for Plastics Van Nostrand Reinhold Company, New York, 1978.

2. Plueddemann, E.P. Silane Coupling Agents Plenum Press, New York, 1982.

3. Monte, S.J. and Sugerman, G. 40th SPE-ANTEC 1985, Paper No. 202.

4. Nakatsuka, T, Kawasaki, H and Itadani K. J. Appl. Polym. Sci. 1982, 27, 259-269.

5. Cohen, L.B. 39th Ann. Tech. Conf. RP/CI. SPI. 1984, Session 19E, 1-5.

6. Schonhorn, H. and Frisch, H.L., J. Polym. Sci. Polym. Phys. Ed., 1973, 11, 1005-1011.

7. Reed, P.E., Impact performance of polymers. In Developments in Polymer Fracture -1, ed E.H. Andrews. Elsevier Applied Science Publishers, London 1979, pp 121-153.

8. Broutman, L.J. and Sahu, S., Mater. Sci. Eng. 1971 8, 98.

9. Trachte, K and DiBenedetto, A., Int.J. Polym. Mater. 1971, 1, 75.

10. Outwater, J.O. and Murphy M.C. Modern Plastics 1970, 7, 160.

11. Cottrell, A.H. Proc. R. Soc. 1963, Ser. A., 282, 1.

12. Copeland, J.R. and Rush, O.W. Plastics Compounding, 1978 November/December, 26-41.

    Copeland, J.R. Good, S and Phillips G.C 38th Ann. Conf. RP/CI. SPI. 1983, Session 13-G, 1-3.

    Power, T. Industrial Minerals, 1986, 19-34.

13. Kelly, A. Proc. Royal Soc., 1964, A282, 65

14. Kelly, A. and Tyson, W.R. J. Mech. Phys. Solids, 1965, 13, 329-350.

15. Darlington, M.W., McGinley, P.L. and Smith, G.R. J. Mater. Sci., 1976, 11, 877.

16. Darlington, M.W. McGinley, P.L., J. Mater. Sci., 1975, 10, 906.

17. Bader, M.G. and Bowyer, W.H. Composites, 1973 July, 150.

18. Nicolais, L., and Narkis, M., Polym. Eng. Sci. 1971, 11, 194-199.

19. Chow, T.S. J. Mater. Sci, 1980, 15, 1873-1888.

20. Nielsen, L.E. J. Appl. Phs 1970, 41, 4626.

21. Nielsen, L.E., Ind. Eng. Chem. Fundam, 1974, 13, 17.

22. Nielsen, L.E. J. Appl. Polym. Sci., 1973, 17, 3819.

23. Darlington, M.W. and Christie, M.A. The Role of the Polymeric Matrix in the Processing and Structural Properties of Composite Materials, ed. J.C. Seferio and L. Nicolais, Plenum Publishing Corporation, 1983 pp 319-355.

24. Darlington, M.W and Wright, P.J. unpublished work.

25. Theocaris, P.S., Colloid Polym. Sci. 1984, 262, 929-938.

26. Theocaris, P.S. and Ponirides, P.H. J. Appl. Polym. Sci. 1986, 32, 6267-6279.

27. Maurer, F.H.J. Kosfeld, R and Uhlenbroich, Th, Colloid, Polym. Sci. 1985, 263, 624-630.

28. Simha, R. Jain, R.K. and Maurer, F.H.J. Rheol. Acta. 1986, 25,161-165.

29. Kendall, K and Sherliker, F.R. Brit. Polym. J., 1980, 12, 85-88.

30. Private communication from Jones, D.V.C., Materials Development Division, AERE, Harwell, Didcot, Oxon.

31. Godlewski, R.E. 38th Ann. Tech. Conf. RP/CI. SPI, 1983 Session 13-E, 1-5.

32. McFarren, G.A. Sanderson, T.F. and Schappell, F.G. Polym. Eng. Sci. 1977, 17, 46-49.

33. Von Mises, R. Gottinger Nachrichten, Maths-Phys. Klasse, 1913, 582.

34. Radosta, J.A. 34th SPE-ANTEC, 1976 465-471.

35. Riddell, M.N. and O'Toole, J.L. 23rd Ann. Tech. Conf. RP/CI, SPI, 1968, Section ZE, 1-5.

ADHESION AND PROTECTIVE PROPERTIES OF POLYAMIDOIMIDE
COATINGS ON COPPER SUBSTRATE

V.M. Startsev, R.Z. Kaz'mina, V.A. Ogarev and V.N. Buryanenko
Department of Polymer Coatings,
Academy of Sciences of the USSR,
Lininsky prospekt, 31, Moscow 117915,
USSR

INTRODUCTION

Electronics and electrical engineering as well as other branches of industry have a great need for lacquers, enamels and coatings possessing durable serviceability over the temperature range 200-220°C. For these purposes, commercially available polyamidoimides (PAI) combining the advantages of polyamides and polyimides, can be used.

At present there is a lot of literature data on PAI, most of it concerning the synthesis, structure and properties of free films [1-6]. The analysis of the processes occurring during the formation of the coatings and the factors affecting these processes, as well as the study of the physico-mechanical and usability properties of the coatings have not been done.

The obligatory condition for the formation of a stable polymer coating is providing a reliable adhesion contact with the substrate. This major process occurs, as a rule, against the background of other processes (chemical reaction, removal of the solvent, destruction etc.). The matters are complicated by the factors affecting the above-mentioned processes: energy characteristics and electrochemical activity of the substrate, preparation of its surface, character of interaction on the interface, supermolecular structure of the polymer, and finally the hardening temperature.

The protective properties of a polymer coating cannot be brought down only to its barrier (diffusion) characteristics.   Metal with a coating is to be considered as a specific electrochemical system, where the corrosion process develops in a step by step way: diffusion of the aggressive medium through the coating, destruction of the adhesion contact, the yield of the corrosion products.

The effect of the chemical structure of polyamidoimides (PAI) on the adhesion and protective properties of coatings on a copper substrate during the process of their hardening was studied.   The analysis was done on the assumption that the corrosion of metals with a polymer coating is a step by step process.   Physico-chemical processes occurring on the interface, and the phase state of the polymer were also taken into account.   The mechanisms of the formation and destruction of the adhesion contact copper-PAI and the role of the chemical and supermolecular structure in these processes were also studied.   The change of the adhesion and protective properties of the coatings in the process of hardening turned out to be considerable. Temperatures ensuring the maximum level of the properties under study are suggested.

EXPERIMENTAL

Materials

Polyamidoimides (brand PAI-1 and PAI-2), whose chemical structures differed only in the type of the bridge groups in the diamine component were used:

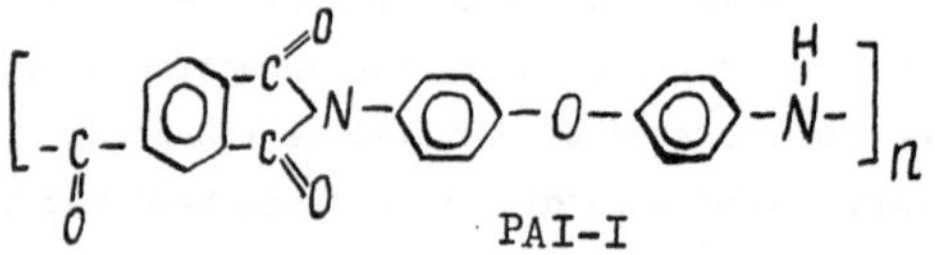

PAI-I

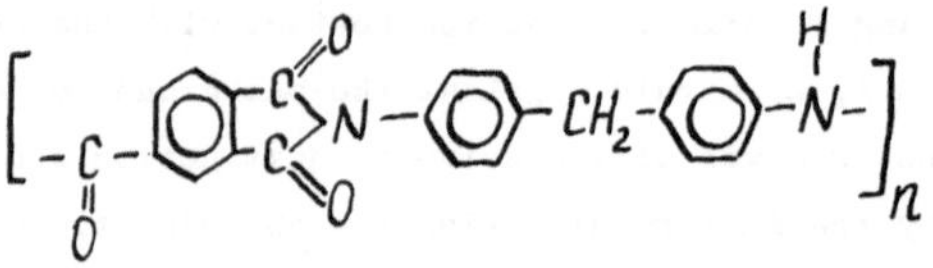

PAI-2

The PAI lacquers were 15% solutions in methylpyrrolidone (N-MP) where for PAI-1, $M_W = 5.2 \times 10^4$; $n_s = 0.74$ and for PAI-2, $M_W = 3.2 \times 10^4$; $n_s = 0.44$ respectively.

A 50 μm thick copper foil was used as a substrate. The specific surface energy of the foil was 43.3×10^{-3} J/m^2 (CuO). The foil surface was degreased by acetone and wiped with N-MP before depositing the PAI solution.

Formation of Coatings

The PAI coatings were deposited in the following way: Glass plates were covered with the foil and the polymer solution was poured on the foil surface to create a 50 μm thick coating. The hardening of the coating was carried out step by step over the temperature range 100-350°C under normal conditions.

Methods

After each hardening stage the coated foil was removed from the glass and cut into strips 5 mm wide. The length of the strips provided a 40 mm working base for peeling off the foil from the polymer films at an angle of 180° at a rate of 17.10^{-5} m/s. The peeling (10 samples after each hardening stage) was carried out by means of the universal test apparatus "INSTRON" (Model 1121) at room temperature. The spread in the average results did not exceed 15%.

For the study of the supermolecular structure of the polymer layers bordering on the foil we used electron microscopy ("Philips", Model EM-301) of the PAI film samples after peeling. The surfaces of the films were prepared by a technique due to Kischer, Evko and Lukianovich [7].

To determine the character of interaction in the pair copper-PAI, we recorded the IR-spectra of the mirror reflection (angle 80°, spectrophotometer Perkin-Elmer, Model 983 G) from the surface of the initial foil and from that after peeling the polymer. Then from the second spectrum we subtracted the first one thus obtaining the difference spectrum. We also recorded the IR-spectrum by MATR (KRS crystal, 45°) of the free film of the polymer.

The protective properties of the PAI coatings were studied by the capacitance-Ohmic method [8]. On 3 mm thick copper plates 70 μm thick coatings were formed over the same hardening temperature range. The experiments were done in the 0.5 N NaCl solution during 6 months. The stability of the protective properties of the coatings was evaluated according to the change of the frequency dependence of the capacitance C and the resistance R of the coloured electrodes in the course of time.

RESULTS AND DISCUSSION

Adhesion Properties

Figure 1 shows the data on the change of the forces of peeling the copper foil from the coatings at different stages of hardening. We observed the closely similar character of the curves for both the PAIs. The adhesiogramms of the peeling process were also identical. These effects are due to the same character of interaction and destruction of the adhesion contacts of both PAIs with the copper foil. The maximum peeling forces were observed after hardening at 140°C (413°K), the PAI-1 coatings having better advantages. The peeling forces appeared to be unexpectedly low over the hardening temperature range 200°-300°, where for free films we observed an increase in the strength.

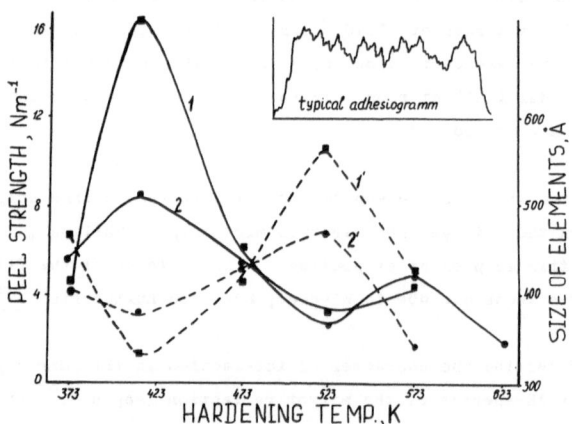

Figure 1.   Influence of the hardening temperature on the value of peeling forces of the polymer coating from the copper foil (1,2) and on the size of structural elements (1', 2'): 1,1' - PAI-1, 2,2' - PAI-2.   A typical adhesiogramm (hardening temperature 200°C).

Figure 2 shows the change of the supermolecular structure of the polymers on the side of the foil at different stages of hardening (after peeling). Both PAIs have a macroheterogeneous structure, which consists of small globular elements. This structural organization is obviously "responsible" for the change of the peeling forces, because the change of the size of structural elements is inversely proportional to the change of the peeling forces (Fig. 1).

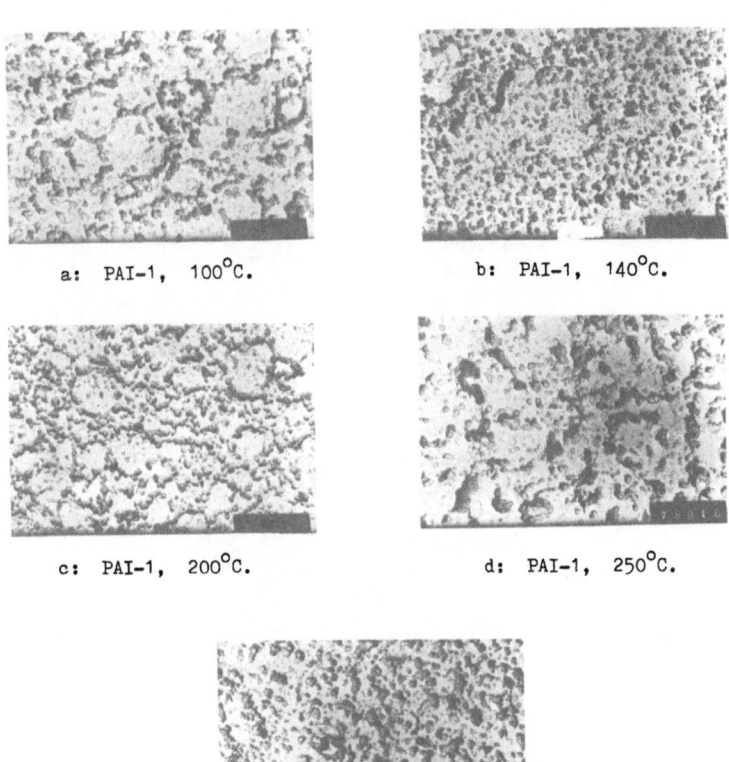

a: PAI-1, 100°C.

b: PAI-1, 140°C.

c: PAI-1, 200°C.

d: PAI-1, 250°C.

e: PAI-1, 300°C.

Figure 2   continued:-

60

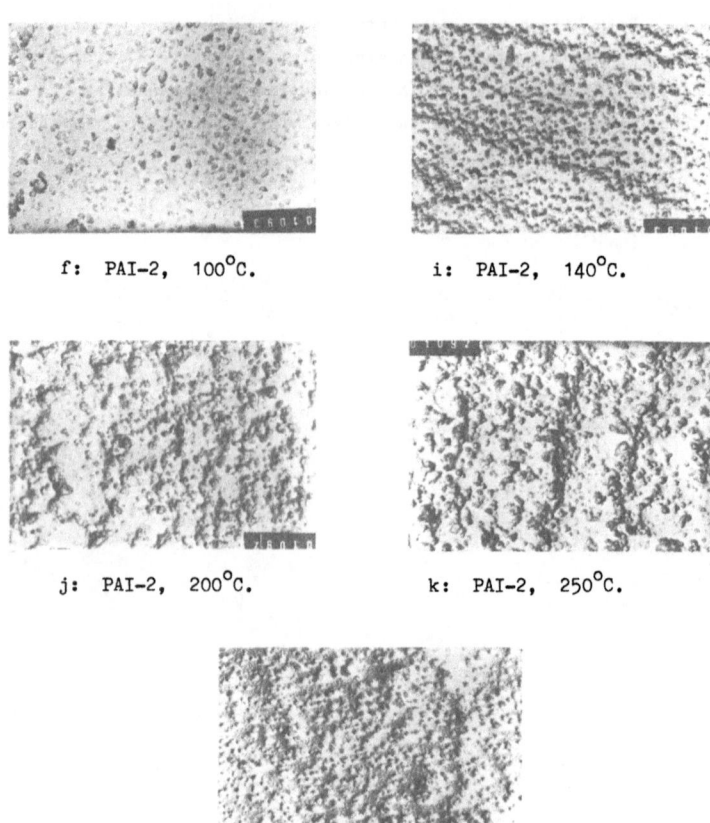

f: PAI-2, 100°C.

i: PAI-2, 140°C.

j: PAI-2, 200°C.

k: PAI-2, 250°C.

l: PAI-2, 300°C.

Figure 2.   Microphotographs of the supermolecular structures
of the polymers on the side of the substrate at
different stages of hardening:

Taking into account the above-mentioned facts, we propose the following mechanism of interaction in the pair copper-PAI. The initial PAI solution is an associative viscous liquid. When the solution is deposited on the foil the structural elements (and not individual macromolecules) of the polymer are adsorbed on the active centres (or their clusters) of the surface and become fixed on them due to the donor-acceptor interaction, with the formation of the chelate complexes [9]. This assumption was tested by means of IR-spectroscopy. In the difference spectrum (Fig. 3),

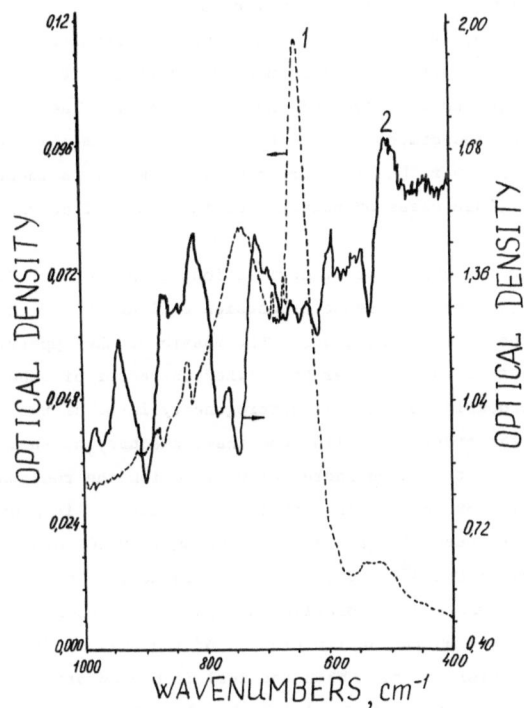

Figure 3. The difference IR-spectrum of the copper foil reflection (1) and the MATR spectrum of the polymer (2) after hardening at 140°C.

intensive bands at 740 and 650 cm^{-1}, which are absent from the polymer
spectrum, are observed.   According to recommendations of Chulkevich et al.,
the band at 650 cm^{-1} can be identified as the Cu-O bond, i.e. for PAI as
the Cu -$\boxed{\text{O=C- N}}$ bond (C=O in the imide cycle).   IR-spectra presented in
Fig. 3 belong to the PAI-1 coating hardened at 140°C.   But there appears
a specific interaction obviously during the initial contact of the PAI solution with the substrate and this is preserved at all the stages of hardening.

The advantages of the pair, copper-PAI-1, as far as the peeling forces are concerned can be explained in the following way.   The elements of
the PAI-1 structure, whose dimensions are smaller, involve a larger number
of the foil active centres into the interaction (the effect analogous to
the increase in the contact area of the components).   The reasons for the
extreme change of the dimensions of the structural elements are not
discussed in this paper, but they are probably similar to those considered
earlier for the close class of polymers on the basis of polyamic acid |11|.

As far as the character of destruction of the contact PAI-copper
foil is concerned, one can assume a cohesion destruction of the polymer in
places of the specific interaction.   The mosaic of the appearing contacts
found its reflection in a complex "saw-like" character of adhesiogrammes.
These latter, probably reflect the heterogeneity level of the distribution
of contacts at the front of peeling and the inadequacy of their energy
characteristics.   At such character of destruction the resistance to
peeling is limited by the strain ability of polymers.   In fact, the
decrease in the rupture elongations from 20-25% to 8-10% over the hardening
temperature range 140-250°C led to a three-fold decline of the peeling
forces (Fig. 1).   The above-described mechanism is presented in Fig. 4.
The polymer is fixed on the active centres of the foil surface as a result
of the specific interaction.   When the foil is peeled off the polymer is
stretched forming tension bars.   At the bar length 1_{cr} the strain ability
of the polymer becomes exhausted (this ability naturally depends on its
physical state, i.e. the hardening temperature) and the bar is broken.   The
stretched matrix polymer and its residue on the foil relax but not to a
full extent, forming point microprofiles.

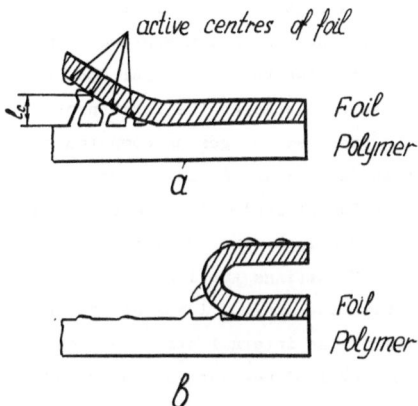

Figure 4.   A simplified model of the destruction of the
adhesion contact during the peeling of the polymer
from the foil: a - initial stage; b - active
process.

Protective Properties

The protective properties of the coatings were determined according
to the cnange of the frequency dependences of the specific capacitance C
and resistance per a cross-section unit R of the coating of coloured copper
plates in the course of time [8].   The stability of the protective properties of the coatings in the corrosion process manifests itself in the
slight change of the capacitance in the presence of a considerable dispersion of resistance over the chosen frequency.

The analysis of the electrochemical data (Fig. 5-8) showed the following.   The barrier properties of PAI-1 and PAI-2 turned out to be close: before the exposure in NaCl solution R = 2 x 10^8 - 2 x 10^9 Ω cm^2 and C = 250-300 pF/cm^2; a regular improvement of these characteristics caused by the rise of the hardening temperature from 140 to 350°C (solid lines in the figures) was observed.   The PAI coatings hardened at 140° and 350°C, after the exposure, displayed the least protective properties: their R and C underwent the greatest changes as compared to the initial values; the frequency dependences of R and C also changed abruptly.   The PAI-1 coatings displayed the lowest protective properties after hardening at 140°C, while the PAI-2 coatings were the least stable after hardening at 350°C. But if for PAI-2 coatings the low level of the protective properties at 350°C correlates with the minimum level of the peeling forces (see Fig.1) and the maximum level of the internal stresses then for PAI-1 coatings we observed the maximum level of peeling forces at 140°C.   Thus, it can be assumed, that it is not the absolute value of the peeling forces that limits the protective properties of the coatings, but the physico-chemical processes occurring in the polymer volume and on the interface under the influence of the aggressive medium.

The point is that the hardening of the solution compositions of polyimides and PAI is accomplished by the competition of a number of processes: the removal of the solvent, imidization (or additional imidization for PAI) and a low-temperature decomposition of polyamic acid [11].   The PAI studied appeared to lose the molecular mass and density of the polymer, when heated from 100 to 140°C.   These processes are accompanied by the crushing of the structural elements (Fig. 2) and plastification of the coating by the solvent released during the crushing of associates.   The plastification causes an increase in the strain ability of the polymer leading to an increase in the peeling forces (Fig. 1).   But the reorganization of the structure and plastification of the volume has a negative effect on the protective properties of the coatings (the "loosening" of the polymer and concentration of the solvent on the interface).

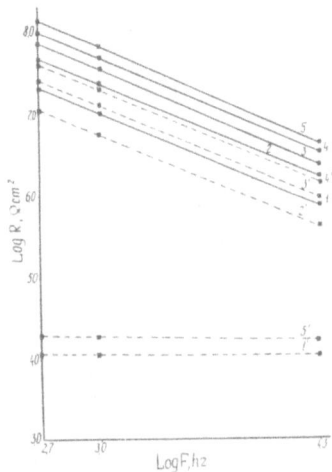

Figure 5.   Frequency dependence of resistance R of the PAI-1
             coatings before (solid line) and after (dotted
             line) the exposure in NaCl at different stages
             of hardening: 1,1' - 140°C; 2,2' - 200°C;
             3,3' - 250°C; 4,4' - 300°C; 5,5' - 350°C.
             The same symbols are used in Figs. 6-8.

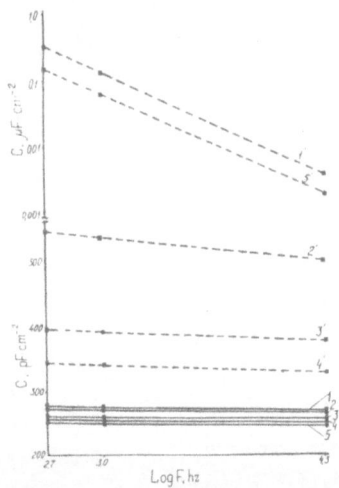

Figure 6.   Frequency dependences of the specific
             capacitance C of the PAI-1 coatings.

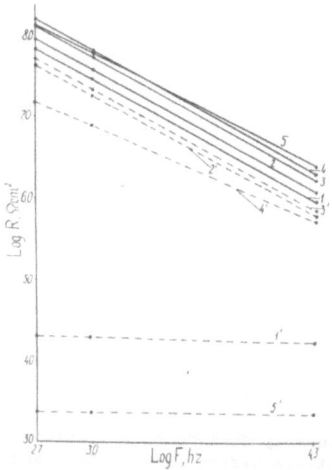

Figure 7.   Frequency dependences of resistanc R of the PAI-2
                 coatings.

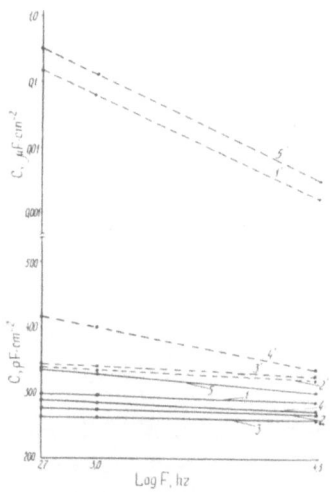

Figure 8.   Frequency dependences of the specific capacitance
                 C of the PAI-2 coatings.

According to the data of Figs. 5 and 6, the highest protective properties of PAI-1 coatings are formed after their hardening over the temperature range 250-300°C: the R and C values of these coatings are subject to the least absolute changes, their frequency dependence remaining stable.

In terms of the same criteria 200-250°C are the optimum hardening temperatures for PAI-2 coatings (Figs. 7 and 8). The upper limits of the above-mentioned hardening temperature ranges provided for the maximum strength and density levels of the respective polymers. Moreover, after the hardening at these temperatures the coatings on the basis of both PAI after the exposure had better and equal values R (2×10^7 Ω cm^2) and C (340 pF/cm^2). This enables us to assume the limiting effect of the polymer density (barrier properties) on the kinetics of the corrosion process.

CONCLUSIONS

The experimental results obtained and their analysis lead us to conclude that the closeness of chemical and supermolecular structures of the PAI under study ensures the closeness of the adhesion and protective properties of the coatings. The difference of the polymers in molecular mass and viscosity of the initial solutions manifests itself over different hardening temperature ranges providing for the maximum level of protective properties (up to 300°C for PAI-1 and up to 250°C for PAI-2). The protective properties as well as the level of peeling forces are limited by the properties of the polymer component, barrier and strain properties, respectively, i.e. finally by the level of intermolecular interaction.

The use of model samples (foil plus polymer coatings) to determine the peeling forces failed to reveal the level of "true" adhesion occurring in coatings on massive plates and obviously correlating with the level of protective properties. And, though the discovery of such correlations was not within the scope of the given study, it opens a way to the oriented regulation of the complex of service properties of the coatings.

68

REFERENCES

1. Chudina, L.I., Litovchenko, S.I., Spirina, T.N. and Chukurov, A.M., Plasti. massy, 1970, N8, S.12.

2. Suvorov, B.V., Zhubanov, B.A., Mashkevich, S.A., Trimellitovaya kislota i polimery na yeo osnove (Alma-Ata, "Nauka", 1975), S.289.

3. Khomenkova, K.K., Zamulina, L.I. and Poliamidoimidy, -V kn.: Novye polymernye materialy (Kiev, "Naukova dumka", 1980), S.162-180.

4. Astakhin, V.V., Trezvov, V.V. and Sukhanov, I.V., Elektroizolyatsion-nye laki (Moskva, "Khimiya", 1981), s.216.

5. Campa, J.G., Abajio, J. and Nieto, J.L., Macromol. Chem., 1982, 183, p.571.

6. Li, G., Stoffi, D. and Nevill, K., Novye lineinye polimery (Moskva, "Khimiya", 1972), s.133-158.

7. Kiselev, M.R., Evko, E.I., Lukianovich, V.M., Zavod. labor., 1966, 32, 201.

8. Rozenfeld, I.L., Burianenko, V.N., Zhigalova, I.A., Lakokras. mater. i ikh primenenie, 1966, N 3, S.62; 1966, N 5, S.55.

9. Entsiklopediya polimerov (Moskva, "Sov. entsiklopediya, 1972), t.1, s.1004.

10. Chulkevich, A.K., Lavrentiev, I.P., Moravskiy, A.P., Khidekel, M.L., Ponomarev, V.I., Filipenko, O.S., Atovmian, L.O., Koordin. khimiya, 1986, t.12, 470.

11. Startsev, V.M., Chugunova, N.F., Morozova, N.I., Nesterov, V.V., Gerasimov, V.D. and Ogarev, V.A., Vysokomolek. soed., 1987, t.29, s.458.

ADHESION OF NITRILE AND ETHYLENE-PROPYLENE RUBBER TO MOULD MATERIALS

R .K. Champaneria, M. Lotfipour, D.E. Packham*,
School of Materials Science,
University of Bath, Bath, BA2 7AY

D. Brister,   and   D.M. Turner,
Victaulic plc,   *Avon Rubber Company,*
Huntingdon, PE18 7DJ   *Melksham, SN12 8AA*

INTRODUCTION

Much of the thrust of experimental work in adhesion is towards the production of stronger or more durable bonds. However the point should not be lost that a major object of a scientific study of adhesion is to gain an understanding of the phenomenon: this can then be exploited to obtain a level of adhesion appropriate to the particular situation.

The need for widely differing levels of adhesion according to the circumstances is shown within the context of rubber to metal adhesion. Rubber-metal bonds form an essential element in many examples of modern vehicle suspensions, bridge bearings and steel-reinforced tyres[1,2]. Here high adhesion for the lifetime of the article is required. By contrast, low but controlled adhesion, is ideally required in rubber moulding. Low so that the moulding can be easily removed, controlled so that when the split mould opens, the moulding always remains in the same half, and can be removed by application of a consistent force.

Problems of mould release and the related phenomenon of mould fouling are widespread. The plethora of descriptions of mould release agents in the patent literature shows that none is really satisfactory. As long ago as 1971 it was estimated that mould fouling cost the British rubber industry alone £500,000 pa[3].

*To whom correspondence should be addressed.

Recent trends in the rubber industry from compression and transfer moulding towards injection moulding with its shorter cure time, but higher cure temperatures, have aggravated the problem. Similarly moves towards automation of the whole moulding operation makes reliable demoulding essential.

A typical rubber compound can easily contain half a dozen to a dozen ingredients. Add to that a wide range of mould alloys, surface finishes and vulcanisation conditions, and it is not surprising that experimental studies have revealed the complexity of the phenomena of mould release and fouling. After a wide-ranging study Maclean and Morrell lamented the "*unpredictable and variable nature of the phenomenon*"[4].

MEASUREMENT

A major problem of work in this area, which contributes to the difficulty of obtaining reproducible results, is that of measurement. The very low levels of adhesion involved have to be measured under conditions sufficiently close to industrial practice for the results to be relevant.

Many of the earlier studies [4,5,6] relied upon some form of counting, for example the number of test discs remaining in a mould after a standard demoulding procedure. Thus a major study at the Laboratoire de Recherche et de Contrôle du Caoutchouc [6] used an injection mould with 32 cavities in which metal test discs could be inserted, figure 1. The level of mould release was indicated by the "percentage adherence" defined as the percentage of the 32 discs remaining in the mould after automatic demoulding.

Direct Measurements

TMS Rheometer. In the present work considerable use has been made of the Turner, Moore and Smith rheometer [7,8]. This highly instrumented rheometer, which is a development of the Mooney rheometer, has a transfer chamber above the traditional Mooney cavity (figure 2). The general lay-out is shown in figure 3.

The rheometer is used with a biconical rotor with a tiny couette region, which means that at a constant rate of rotation, the vast majority of the rubber is subject to a constant shear stress [9].

In order to study mould release the rheometer is used as follows [10]. A rotor of the mould alloy of interest is inserted and a blank of the rubber is placed in the transfer cavity. This is then injected around the stationary rotor, and cured at the appropriate temperature for a predetermiend time. The motor is started and the peak shear stress required for the rotor to break free from the rubber (figure 4), termed the "mould sticking index", is taken as a measure of mould sticking. Full details are given elsewhere [10].

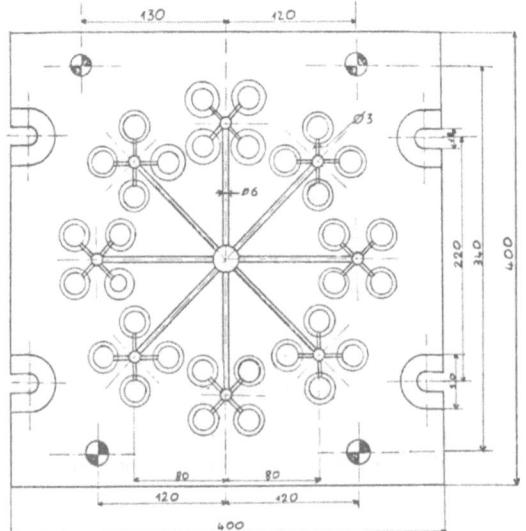

Figure 1. 32 cavity mould used by LRCC[6]

Blister Test. The blister test has been used as a measure of adhesion of a polymer to a rigid substrate by many authors [11-15]. A disc of polymer is applied to the substrate with a central penny-shaped crack between polymer and substrate. To measure the adhesion the crack is pressurised (eg by applying nitrogen through a hole in the centre of the substrate) until a critical pressure P_c is reached at which the crack propagates at, or close to, the interface (figure 5).

Briscoe and Panesar [16,17] have applied the technique to the release of polyurethane rubber from steel, and have successfully studied the action of various types of release agent. The blister test gives an "energy of fracture" G for the polymer-metal system, obtained from computing the stored energy in the deformed polymer at the initiation of failure.

There are several mathematical treatments of the blister test in the literature [18] for polymer layers of various thicknesses. They differ partly according to the relative contributions considered of bending and stretching energy and partly according to the degree of approximation employed. Briscoe and Panesar favoured for their system an

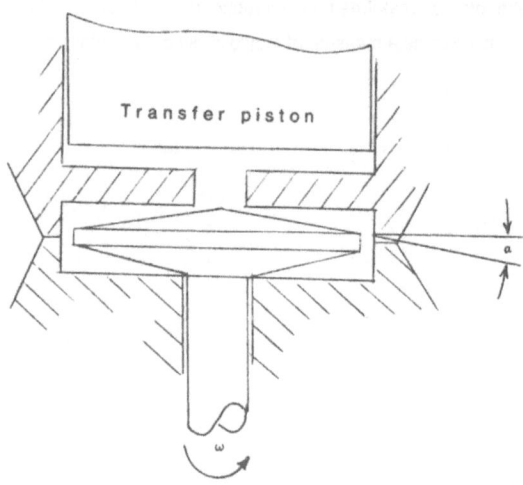

Figure 2.   TMS rheometer. The injection ram, transfer chamber and rotor cavity with rotor.

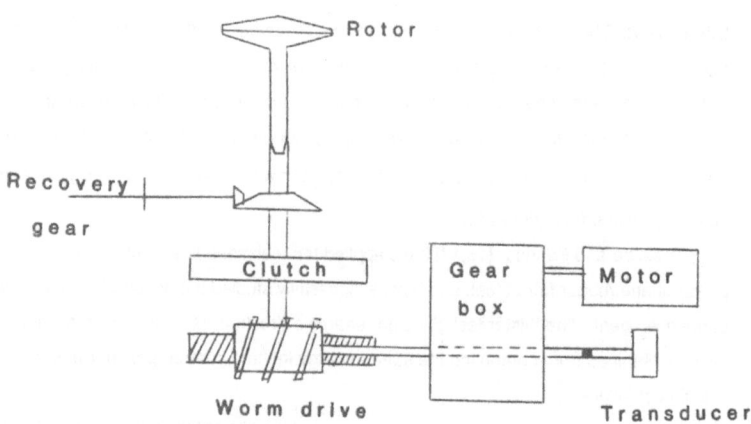

Figure 3.   Simplified diagram of the TMS rheometer

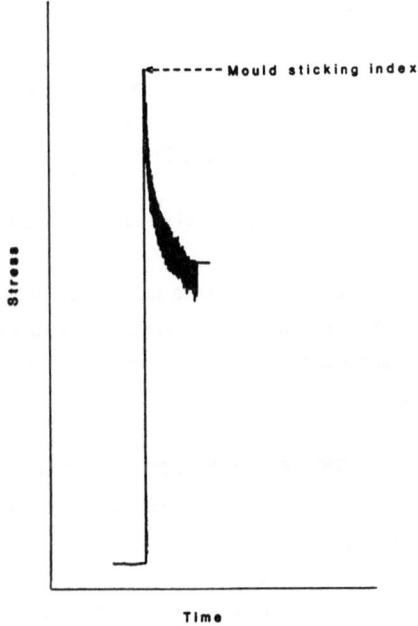

Figure 4.   Result of a typical mould sticking experiment.

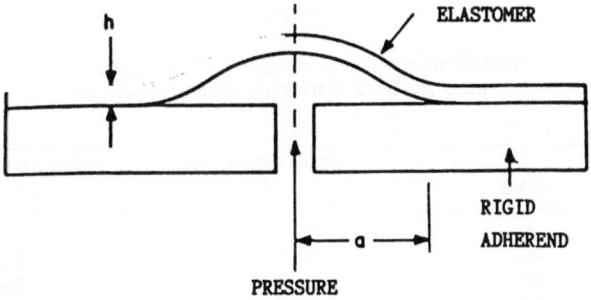

Figure 5.   Configuration of the blister test, a is the debond radius, h the thickness of rubber.

approach which treats the rubber as a stretched membrane and gives [17,18]:

$$G = \left[\frac{P_c^4 \, a^4}{576 \, Eh \, (1 - v^2)} \right]^{1/3}$$

where a is the debond radius, h the thickness of the rubber, and E and v respectively its Young's modulus and Poisson's ratio.

The present authors have adapted Briscoe and Panesar's technique to study mould release of conventional heat-cured rubbers by developing a mould which enables the rubber discs to be injection moulded. Preliminary results (table 1) show results similar to those from the TMS rheometer.

INFLUENCES ON MOULD STICKING

One of the reasons for adopting the blister test in the present work is to enable the effect of large numbers of injection cycles to be ascertained. This has not yet been done. However the LRCC study [6] found a quasi-periodic variation of adherence with cycle number (figure 6). With a new mould there was first a fall in adherence to a "plateau" value constituting the working phase of the mould. After a number of cycles (ideally a large number) a significant deposit begins to form on the mould, and the adherence increases. After a while the deposit tends to break off, leading to a drop in adherence and a second phase. The TMS is only suitable for studying small numbers of cycles, so the results discussed below refer to the early part of the cycle.

TABLE 1

Preliminary comparsion of results for TMS mould sticking
index and blister test fracture energy for En8 steel

Rubber		Mould sticking index	Fracture energy
Type	Code	kPa	J/m^2
NBR	NI	117	0.72
NBR	N2	182	0.87
EPDM	EP3	340	1.0

Apart from difficulties of measurement, mentioned above, another problem in

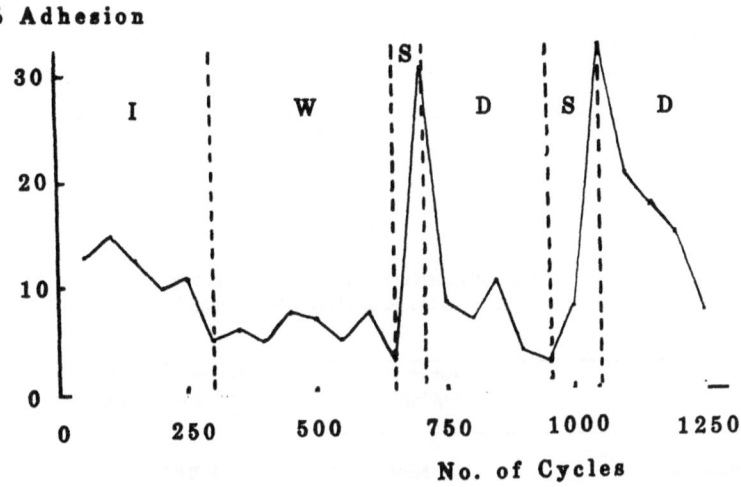

Figure 6. Typical pattern of mould release found by L.R.C.C. (6). I-initiation; W - working phase; S - sticking; D - debonding of deposit.

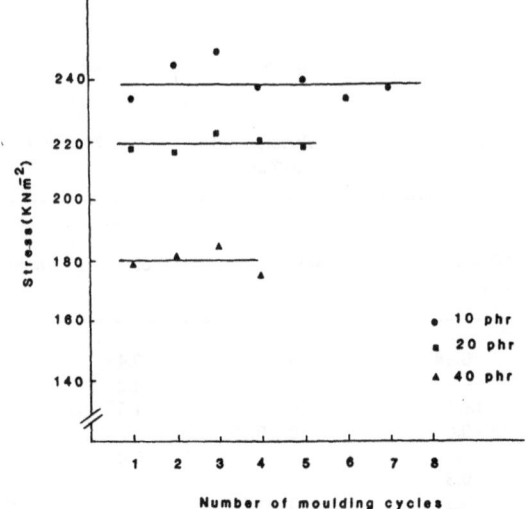

Figure 7. Influence of processing oil on sticking of E.P.D.M. rubber to a KE 355 rotor.

work in this area has been the enormous number of potential variables. A typical rubber formulation contains a large number of types of ingredients - filler, curing agents, stabilisers etc - of which the chemical nature and concentration may be varied. Then of course for any specific rubber compound the cure conditions and mould surfaces may also be varied.

TABLE 2

Materials used in the mould sticking studies
A rubbers; B mould steels (Wt. %: balance Fe)

A	Nitrile Rubber	Ethylene-propylene-diene rubber	
Abbreviation	NBR	EPDM	
Comonomer composition (mole %)	butadiene 72 acrylonitrile 28	ethylene 55.5 propylene 40 ethyldiene norbornene 4.5	
Mooney viscosity ML+4 125°C	45-50	46	
Cure system	Accelerated sulphur	Peroxide	

B	Stavax 420	En8	KE355	Impax
C	0.38	0.4	0.43	0.36
Ni	-	-	4.2	1.4
Cr	13.6	-	1.5	1.4
Si	0.8	0.5-0.35		0.3
Mo	-	-	0.22	0.3
V	0.3	-	-	-
Mn	0.5	0.6-1.0	-	0.2
S		0.06		
P		0.06		

In order to keep the variables to a minimum, two polymers, accelerated sulphur-cured NBR and peroxide-cured EPDM, have been selected for this project and "standard" formulations for each have been developed. A limited range of mould alloys has been used (table 2). Departures from the standard formulation have been made only when necessary to study particular influences. This approach has led to the identification of *three factors* influencing mould release: the compounding ingredients, the base polymer and mould alloy.

Compounding Ingredients. The mould sticking index can often be altered by the addition of or by changing the concentration of certain compounding ingredients. Thus figure 7 shows a fall in mould sticking index produced by addition of processing oil and table 3 shows the effect on a similar polymer of increasing stearic acid concentration. Of course, altering the compound is a common recourse of the technologist as an *ad hoc* response to mould sticking problems. The TMS rheometer enables the effect on sticking of additive changes to be readily assessed. There are however situations where a change of compound formulation cannot be tolerated.

Base Polymer. In a number of instances with both NBR and EPDM, changes in base polymer from the product of one manufacturer to that of another (always keeping the specification the same), has been found to produce a marked change in mould sticking index. Figure 8 illustrates this for N1 and N2, both similar NBR compositions. The two corresponding base polymers would normally be regarded as interchangeable.

TABLE 3

Effect of stearic acid concentration in experimental EPDM based rubbers on the mould sticking values for a KE355 rotor at 1 rpm.

	EP1	EP2	EP3	EP4
Stearic acid (phr)	0.25	0.5	1.5	3.0
Mould sticking index(kN/m^2)	>600	497	338	145

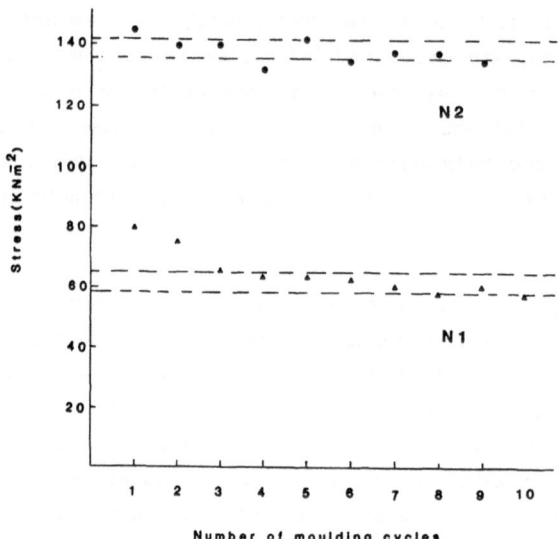

Figure 8. Influence of base polymer on mould sticking; N1 and N2 are identical N.B.R. formulations differing only in the manufacturer, but not the specification of base polymer.

TABLE 4

Mould sticking indices
kN/m^2

Alloy	Nitrile Rubber	
	N1	N2
Stavax-420	98	153
En 8	117	182

Mould alloy. The third effect is illustrated in table 4 where mould sticking indices are given for the same two nitrile compounds N1 and N2 moulded against a carbon steel (En8) and a chromium containing steel (Stavax 420). This effect is of some industrial significance as it provides a way of controlling release without changing the rubber composition in any way and without the use of external release agents. It opens up the possibility of obtaining different release propeties in various parts of a mould.

Surface Analysis.

In order to approach an understanding of the mechanisms which control release, the surfaces involved have been analysed by x-ray photoelectron spectroscopy (XPS). Some of the general features found are briefly discussed here.

Table 5 gives the atomic percentages of the principal elements present in the nitrile compound N1, mentioned above, first based on the compound formulation, and then, by contrast, as found by XPS on the surface after moulding against an En8 TMS rotor. (Hydrogen is ignored as it cannot be detected by XPS). Less nitrogen (principally present in the base nitrile polymer itself) is present on the surface, but the proportion of oxygen, sulphur and zinc are higher than in the bulk. Thus it appears that the moulded surface of the rubber is covered with a layer of non-rubber materials.

TABLE 5

Compositon of NBR compound N1 (a) according to its formulation;
(b) from XPS of the demoulded rubber surface (from EN8)
(Atom %; H ignored)

	(a) Formulation	(b) Surface after moulding
C	93.4	86.5
N	3.9	1.6
O	2.1	8.0
S	0.37	2.3
Zn	0.22	1.0

The counter surface - that of the alloy - is similar. This time it consists of an oxidised metal surface covered with a similar layer to that described above. It is within this interlayer, which forms during moulding, that separation takes place.

Some points can be made about the origin of this layer. Zinc is added to the NBR formulation as zinc oxide, but the binding energy of the zinc found on the rubber surface corresponds to zinc in a compound such as stearate. Stearic acid is also a constituent of the formulation and its reaction product with zinc oxide is found to be one of the compounds which can concentrate at the surface, and affect mould release.

The sulphur binding energy also gives valuable information. Sulphur would be expected to be present as crosslinks ($-C-S_x-C-$) and accelerator residues ($-C-S-$), but much of the sulphur on the surface is in an environment such as that of an (organic) sulphate or sulphonate. Such anions are used as emulsifiers in the polymerisation of acrylonitrile-butadiene. The importance of emulsifier residues is further illustrated by the results in table 6, where quite different mould sticking indices for NBR compounds N2, N3 and N4 are given. These compounds are all of the "standard formulation" and the base NBRs are produced by the same manufacturer in the same way, except that the emulsifiers used during polymerisation were different.

TABLE 6

The influence of residues in the base polymer on the mould release. Mould sticking indices for an Impax rotor with similar NBR formulations. N2 is based upon a commercial rubber, N3 and N4 on pilot plant materials differing only in the emulsifier used.

NBR formulation	Mould sticking index kN/m^2
N2	200
N3	100
N4	50

CONCLUSIONS

In conclusion the notorious "unpredictability" [4] of mould release phenomena is lessening. The TMS rheometer provides a consistent measure of release with acceptable scatter [10]. The blister test shows promise as a good complementary method of assessment. The need to be alive to changes in release properties following changes in compound formulation and base polymer have been shown. Sometimes such changes occur unexpectedly with the change of batch of a previously satisfactory component. The release properties are dominated by the components which come to the rubber-metal interface during vulcanisation. The influence of residues and additives on this process is now becoming clearer.

81

Acknowledgements

The support of the Polymer Engineering Group and SERC is gratefully acknowledged.

REFERENCES

1. Van Ooij, W.J., *Rubber Chem. Techn.*, **51**(1), 52 (1978).
2. Haemers, G., *Adhesion* **4**, 175 (1980).
3. Flemming, C.L., *RAPRA Bull.* p.29 (1971).
4. Maclean, A. and Morrell, S.H., *RAPRA Members' Report* No.7, 1977.
5. Kandazoglou, J., Institut Français de Caoutchouc, 1976.
6. Anon., Laboratoire de Recherches et de Contrôle du Caoutchouc, *Rapport Téchnique* No.120 (1979).
7. Turner, D.M. and Moore, M.D., *Plast. Rubber Processing* 81 (1980).
8. Freakley, P.K., *Rubber Processing and Production Organisation*, Plenum, 1985, p.23.
9. Dinsdale, A. and Moore F., *Viscosity and its Measurement*, Chapman and Hall, 1962.
10. Champaneria, R.K., Harris, B., Lotfipour, M., Packham, D.E. and Turner, D.M., *in press.*
11. Dannenberg, H., *J. Appl. Poly. Sci.* **5**, 125, (1961).
12. Williams, M.L., *J. Adhesion* **4**, 307 (1972).
13. Anderson, G.P., Bennett, S.J., and DeVries K.L., *Analysis and Testing of Adhesive Bonds*, Academic Press, 1977, p.58.
14. Andrews, E.H. and Stevenson, A., *J. Materials Sci* .**13**, 1680, (1978).
15. Stevenson, A. and Andrews, E.H., *Adhesion* **3**, 81 (1979).
16. Briscoe, B.J. and Panesar, S.S., *J .Phys . D .Appl . Phys* . **19**, 841 (1986).
17. Panesar, S.S., PhD thesis, Imperial College, London 1986.
18. Briscoe, B.J., Panesar, S.S., and Kendal, K., *J. Adhesion Sci . Technol*, in press.

IMPROVEMENT AND DIVERSIFICATION OF CYANOACRYLATE ADHESIVES

Dimitar L. Kotzev and Vladimir S. Kabaivanov
Scientific Industrial Centre for Special Polymers
Kliment Ohridski St. 4A, 1156 Sofia, BULGARIA

Cyanoacrylates are enjoying increased popularity as instant-setting specialty adhesives in industrial and consumer markets [1]. Their peculiar properties distinguish them from all other adhesives. They are one-component, catalyst-free and cure in seconds at room temperature. They have exceptionally high adhesion towards most materials – metals, rubber, plastics, wood, ceramics, glass and even live tissue. What is the reason behind these unique properties?

$$CH_2{=}C-C-OR \quad \xrightleftharpoons{Base} \quad CH_2{=}\overset{\oplus}{C}-\overset{\ominus}{C}-OR$$

This is the chemical structure of cyanoacrylate monomer, which is the active ingredient of cyanoacrylate adhesives. R is usually an alkyl radical but could be an alkenyl, an alkinyl or an alkoxyalkyl radical. The combined electromeric effects of the nitrile and alkoxycarbonyl groups can cause exceptionally high polarization of the double bond in the presence of weak bases [2]. Such polarization is also caused by water, which turns out to be the major initiator of cyanoacrylate monomer polymerization. Even following high temperature and vacuum treatment of metal surfaces they remain hydrated and easily adsorb and retain water molecules by hydrogen bond formation. Thus the water adsorbed on all surfaces is the cause for the universal adhesion of cyanoacrylate adhesives.

$$CH_2=\overset{CN}{\underset{}{C}}-COOR \;\rightleftharpoons\; \overset{CN}{\underset{\delta^+\;\;\;\delta^-}{CH_2-\overset{}{C}-COOR}} \;\;\;\;\;\; A^{\ominus} \longrightarrow$$

$$A-CH_2-\overset{CN}{\underset{\ominus}{C}}-COOR \;\;\xrightarrow{CH_2=\overset{CN}{C}-COOR}\;\; A-CH_2-\overset{CN}{\underset{COOR}{C}}-CH_2-\overset{CN}{\underset{\ominus}{C}}-COOR$$

$$\underline{\hspace{3cm}} \;\; -\; polymer$$

The monomer undergoes anionic polymerization at a high rate, producing a polymeric adhesive bond with molecular weight around 600 000 [3].

From the many schemes for cyanoacrylate synthesis, probably the most popular is the following [2]:

$$\overset{CN}{\underset{COOH}{\underset{|}{CH_2}}} \;+\; ROH \longrightarrow \overset{CN}{\underset{COOR}{\underset{|}{CH_2}}} \;+\; H_2O$$

In the first step, cyanoacetic acid is esterified with alcohol. The reaction is catalyzed by organic or inorganic acids. The water formed is removed by azeotropic distillation.

$$n\;\overset{CN}{\underset{COOR}{\underset{|}{CH_2}}} \;+\; n\;HCHO \longrightarrow \left(\overset{CN}{\underset{COOR}{\underset{|}{C}}}-CH_2\right)_n \;+\; n\;H_2O$$

In the next step the distilled cyanoacetate is condensed with formaldehyde in the presence of basic catalyst and dehydrating agents. Rooney has shown that this reaction follows an anionic polymerization mechanism [4].

$$\left\{ \begin{array}{c} CN \\ | \\ C \\ | \\ COOR \end{array} - CH_2 \right\}_n \longrightarrow n \begin{array}{c} CN \\ | \\ C = CH_2 \\ | \\ COOR \end{array}$$

The third step of cyanoacrylate synthesis is vacuum depolymerization of the condensation product.

The purification and stabilization of the monomer is of major importance. As a result of their high reactivity, cyanoacrylate monomers can undergo spontaneous polymerization in bulk, caused by traces of basic residues or water in the monomer or by the influence of room temperature or UV light. Effective stabilizers against radical and anionic polymerization were found to be hydroquinone [5], anthraquinone, sultone compounds [6,7], SO_2 [5], organic sulfonic acids, alkyl sulfates, mercaptanes [8], etc. Our recent studies [9] produced the qualitative dependence of the stability of ethyl-2-cyanoacrylate monomer on the content of water, hydroquinone, SO_2, and methanesulfonic acid, using accelerated testing at $70^{\circ}C$. Similar results are obtained (Fig. 1) for the stability of ethyl 2-cyanoacrylate stored at room temperature. It can be seen that lowering the water content causes exponential increase of the shelf-life. This observation clearly indicates the great influence of water on the stability of cyanoacrylate adhesives. The obtained results are in clear support of the Coover and McIntire claim that adhesives containing less than 0.01% H_2O are stable for over 2 years, while water content above 0.06% causes polymerization [10]. The results for hydroquinone, SO_2 and methansulfonic acid in the range studied, chosen so that the setting time of the adhesive was not affected - 5 to 10 sec for steel, show that the bulk stability increases proportionally with the content of stabilizer.

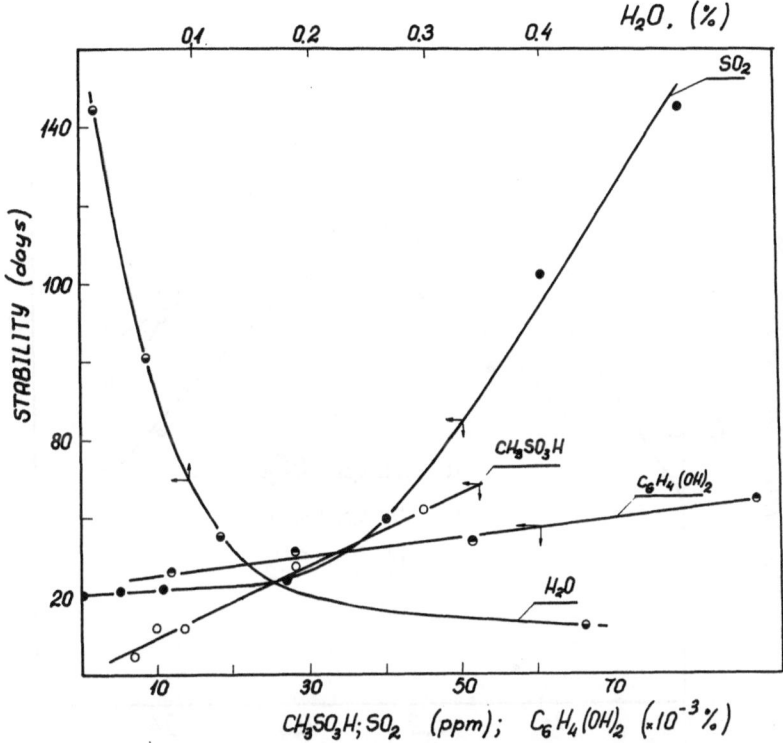

Figure 1.   Dependence of ethyl 2-cyanoacrylate stability
on the content of H_2O, $C_6H_4(OH)_2$, SO_2 and CH_3SO_3H

Possibly due to their universal adhesion and the high strength of the
resultant adhesive bond, only a few studies have been reported in the literature regarding cyanoacrylate adhesion promotion.   Adhesion promoting
action is attributed to the anhydrides of maleic and itaconic acids [11] and
to the esters of gallic acid [12].   Shoenberg describes such properties of
the monocarboxylic acids [13].   More detailed studies on the action of
acetic acid (Fig. 2) reveal that the optimum quantity is in the range of
0.25% acetic acid content [14].   The lap shear strength of bonded soft
steel coupons, which were not chemically treated or activated was measured.

86

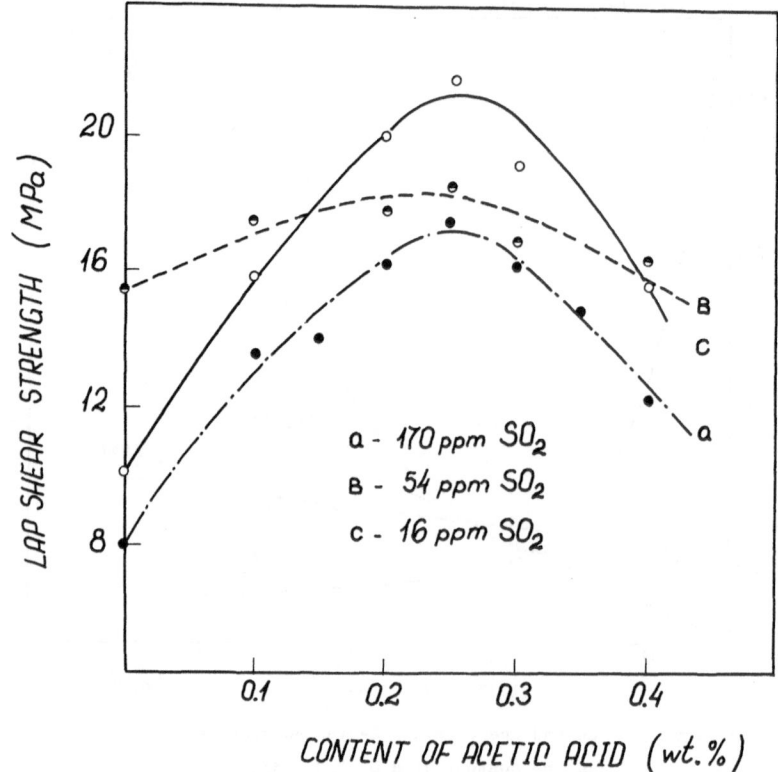

Figure 2.   Dependence of lap shear strength of ethyl
             2-cyanoacrylate bonded joints on the content
             of acetic acid promoter [14].

    Table 1 shows the results of bonding different metal substrates with
and without chemical treatment of the surface [14].   Obviously acetic acid
promotes the adhesion of ethyl 2-cyanoacrylate to steel, stainless steel
and duraluminium but has detrimental effects for copper and brass.   Of
importance is the surface preparation of the specimens.   Best results for
steel and stainless steel are obtained when the surface is degreased only,
but not chemically treated.   Etching, on the other hand, increases by 60%
the lap shear strength of bonded duraluminium samples.   These results can be
explained with the isoelectric point of the oxide surface of steel and

aluminium, which permits the acetic acid molecules to compete with and displace the water molecules on the surface and thus to form strong hydrogen bonds with the oxide [15]. This leads to the assumption that, through the acetic acid intermediate, the cyanoacrylate polar groups can come to a closer contact with the oxide layer, which improves adhesion. Recent results of Suetaka [16] can help the explanation of the dependence of

TABLE 1

The effect of adhesion promotion, substrate nature and surface treatment on lap shear strength of joints bonded eith ethyl-2-cyanoacrylate (ECA) [14]

Type of Substrate	Chemical Treatment (DIN 53281)	Lap Shear Strength (MPa)	
		ECA	ECA contg. 0.25% acetic acid
Steel	no	9.7	19.1
	yes	17.1	6.8
Stainless Steel	no	11.3	18.0
	yes	23.1	6.3
Copper	no	15.1	10.0
	yes	6.4	5.8
Brass	no	12.2	6.7
	yes	2.3	2.6
Duraluminium	no	7.8	8.4
	yes	6.4	13.6

adhesion promotion of acetic acid on the type of surface treatment. Infrared studies of thin cyanoacrylate films on aluminium surfaces have shown hydrogen bond formation between the hydroxyl groups attached to the oxide and the carbonyl oxygen atom in the cyanoacrylate molecule. It has been found furthermore, that the structure of the oxide, changed through various chemical treatment of the metal surface, determines the orientation of the hydroxyl group and therefore the orientation of the carbonyl group of the cyanoacrylate molecule. The actual adhesive bond strength correlates with

the carbonyl group orientation.   Based on these observations, it can safely
be assumed that when a small amount of acetic acid is contained in the cyanoacrylate adhesive it plays a role in the hydrogen bond.formation between the
oxide and the adhesive, possibly changing the geometry and strength of the
interaction.   This can positively or detrimentally affect the molecular
orientation in the polymeric adhesive layer, influencing its mechanical
strength.

One of the properties of a cyanoacrylate adhesive bond which imposes
a limitation on practical utility is the low heat resistance (80°C) which is
due both to the low glass-transition temperature of the polymer and to the
onset of thermal degradation [2].   In an attempt to cope with that problem
we synthesized some cyanoacrylates containing an unsaturated bond in the
ester radical of the molecule [17].   The motivating assumption was that
after typical anionic curing of the monomer, the adhesive bond would be able
to undergo subsequent heat-initiated crosslinking, thus yielding a three-
dimensional structure with improved thermal properties.   The synthesized
alkenyl and alkinyl 2-cyanoacrylates possess the typical adhesive properties
of the alkyl cyanoacrylates, the setting time being 10 to 60 sec.   They may
be ordered in the following sequence with respect to the shear tensile
strength characteristics of steel-steel bonded joints:

$$CH{\equiv}C-CH_2- \quad > \quad CH_2{=}CH-CH_2- \quad > \quad CH{\equiv}C-\overset{\overset{\displaystyle CH_3}{\displaystyle |}}{C}H- \quad > \quad CH{\equiv}C-\overset{\overset{\displaystyle CH_2CH_2CH_3}{\displaystyle |}}{C}H-$$

$$\text{(16.7 MPa)} \qquad \text{(12.4 MPa)} \qquad \text{(11.4 MPa)} \qquad \text{(3.3 MPa)}$$

The relationship is similar to that of the alkyl 2-cyanoacrylates,
i.e. the bond strength depends primarily on the number of C atoms in the
ester group of the molecule:

methyl > ethyl > propyl > butyl > hexyl > heptyl
 (18.0)   (15.6)   (9.3)   (8.0)   (3.1)   (0.9)

The introduction of triple and double bonds into the ester group
increases the tensile strength properties of the cyanoacrylate adhesives.
On the other hand if we try to correlate the thermal resistance of the
adhesives using the data for tensile shear strength of steel-steel bonded
joints treated for 24 h at 150°C we arrive at the following relationship:

$$CH_2=CHCH_2- \quad > \quad CH\equiv CCH_2- \quad > \quad CH\equiv CCH- \quad > \quad CH\equiv CCH- \quad > \quad alkyl$$

with $\overset{|}{C}H_3$ above the third group and $\overset{|}{C}H_2CH_2CH_3$ above the fourth group.

$$(5.0 \text{ MPa}) \qquad (4.0 \text{ MPa}) \qquad (0.9 \text{ MPa}) \quad (0.5 \text{ MPa}) \qquad (0.0 \text{ MPa})$$

Even an adhesive layer based on unsaturated cyanoacrylate with 5 or 6 C-atoms in the ester group has better heat resistance than that of methyl and ethyl 2-cyanoacrylates, whose bonds break after exposure for 1 to 3 h at 150°C. The best results were obtained for allyl and propargyl 2-cyanoacrylates [17].

TABLE 2

Changes of T_g and softening point of cyanoacrylate adhesive bond after 24 h thermal treatment [18]

2-Cyanoacrylate	T_g (°C)			Softening point (°C)		
	20°C	100°C	150°C	20°C	100°C	150°C
Ethyl	101	101	–	80	81	–
Allyl	98	123	154	105	117	180

An argument in favour of structural crosslinking is the change of glass-transition temperature and softening temperature of the adhesive layer after thermal treatment (Table 2). T_g of allyl 2-cyanoacrylate adhesive increases, while that of ethyl 2-cyanoacrylate does not change. Interesting results were obtained on completion of series of non-destructive dynamic shear mechanical tests [19] on cyanoacrylate bonds. The dependence of the reduced shear storage modulus and tan for ethyl 2-cyanoacrylate on temperature (Fig. 3) clearly shows a glass-transition point, a short rubbery plateau, and finally rubbery-liquid flow. The glass-transition related decrease of the modulus and the short rubbery plateau for the allyl 2-cyanoacrylate adhesive bond (Fig. 4) is followed, on the other hand, by a modulus increase proportional to the temperature and characteristic of a rubbery network. Obviously the increased mobility of the polymer chains above T_g along with the available thermal energy produces crosslinking reactions, changing the structure of the adhesive layer and its dynamic

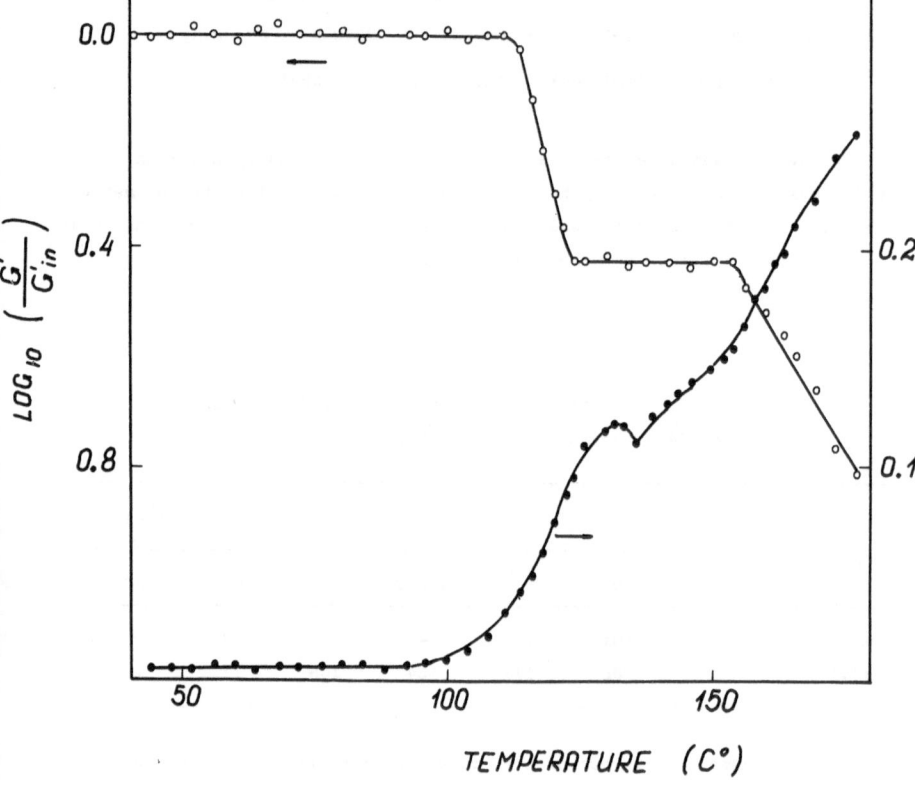

Figure 3.   Dynamic mechanical response of ethyl 2-
              cyanoacrylate adhesive bond [19].

mechanical response.   The dynamic mechanical aprameters of 2llyl 2-cyano-
acrylate aged for 24 h at 100°C prior to testing also indicate cross-linking
induced by the conditions of the experiment.   However, this ageing of the
adhesive prior to testing has produced a shift of the damping peak to higher
temperature and reduced its intensity.   Both of these changes are typical
of crosslinking.   Ageing for 24 h at 150°C produced a polymer having dynamic mechanical response characteristics of a highly crosslinked three-
dimensional structure 19 .   A change in the adhesive bond structure was
observed using SEM photomicrography of fractured steel-steel joints having
different thermal history.   The principal failure mode for ethyl 2-cyano-
acrylate (Fig. 5) is interfacial with little plastic deformation.   At 150°C

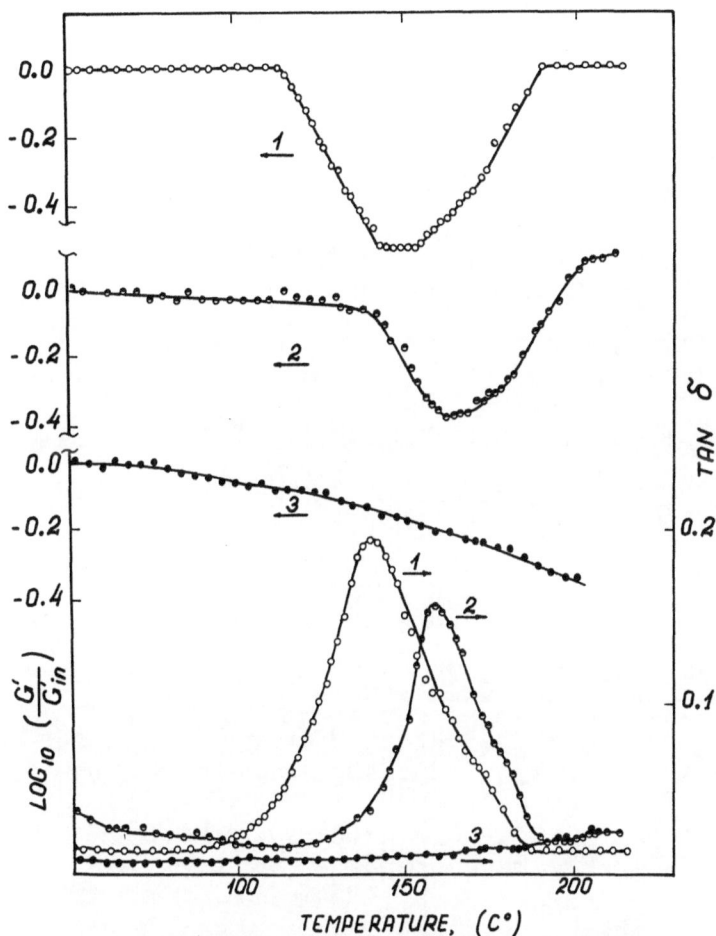

Figure 4. Dynamic mechanical response of allyl 2-cyano-
acrylate adhesive bond after thermal ageing:
(1) 20°C; (2) 24 h at 100°C; (3) 24 h at 150°C
[19].

delamination due to melting of the adhesive occurred at negligible stress.
Fractured specimens of the unheated allyl 2-cyanoacrylate (Fig. 6) have
SEMs that look very similar to those of ethyl 2-cyanoacrylate. Similar
fracture modes are suggested. Ageing allyl 2-cyanoacrylate bonds at 100°C,

Figure 5.   SEM photomicrographs of failed surface of ethyl-
2-cyanoacrylate adhesive joints: (a) room
temperature; (b) after 24 h at 100°C; (c) after
24 h at 150°C; [19] (Mark represents 50 μm).

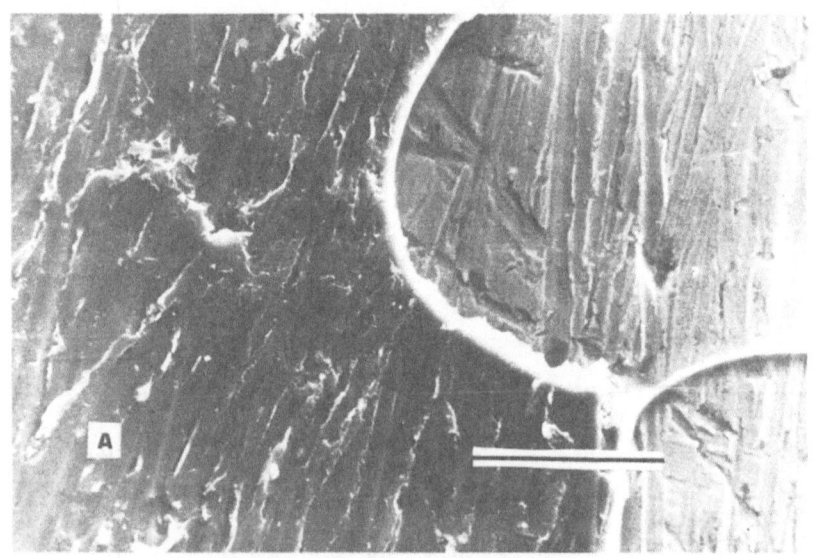

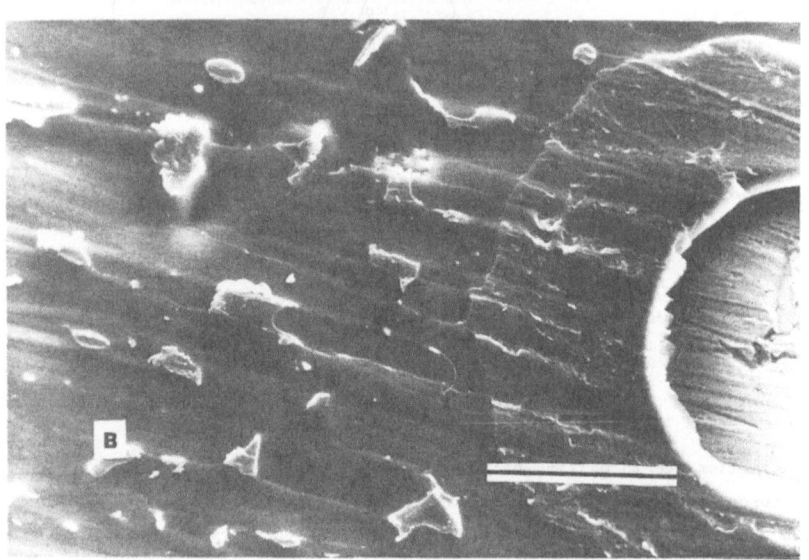

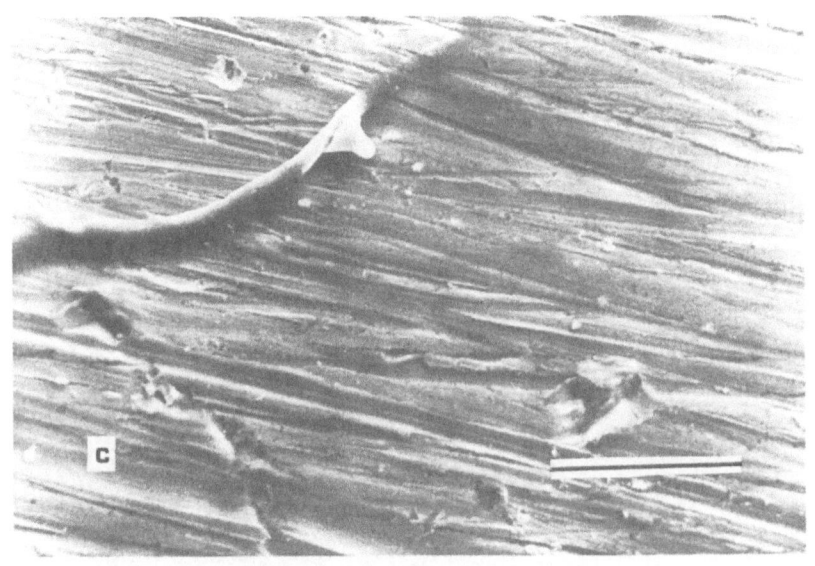

Figure 6.   SEM photomicrograph of failed surface of allyl
             2-cyanocarylate adhesive joints:
             (a) room temperature; (b) after 24 h at 100°C;
             (c) after 24 h at 150°C   [19] .
             (Mark represents   50 μm)

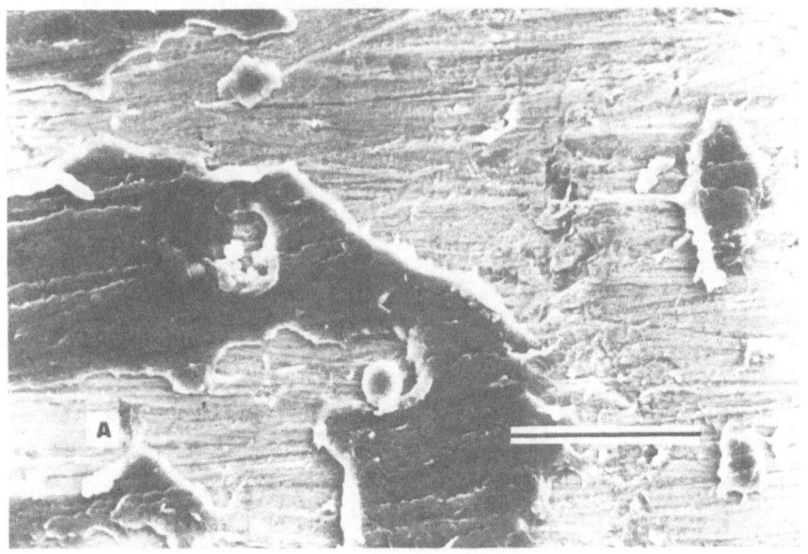

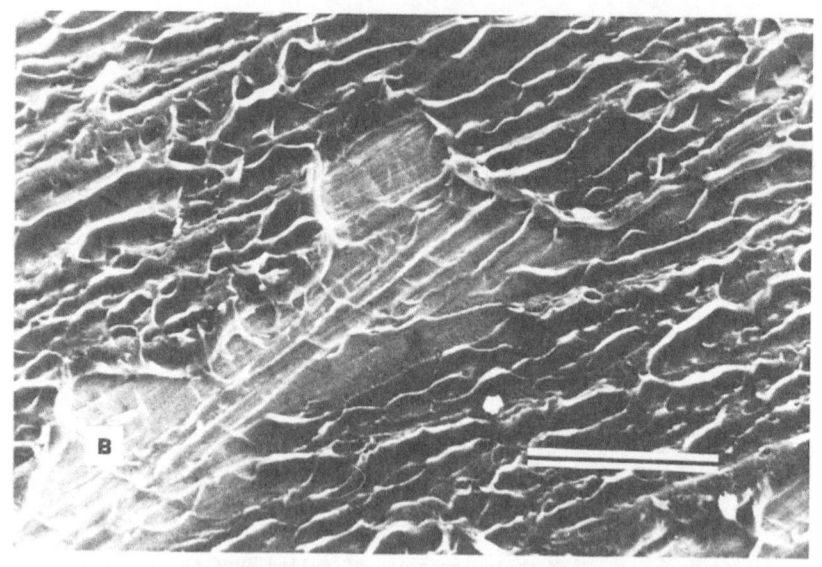

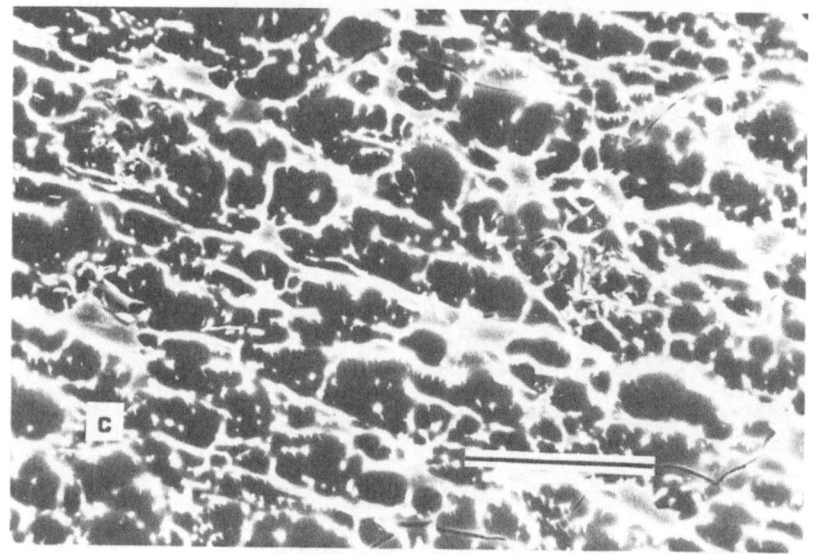

however, brings about a dramatic change in the irreversible micromechanisms in the failure process.   Ridges of plastic deformation may be noticed. The crosslinking process has produced much more resilient-like failure mode. The stress concentrations at the tops of the finish scratches probably initiated plastic mechanisms that absorbed significant amounts of strain energy, leading to increased breaking strength.   Ageing at 150°C induces brittleness of the adhesive bond due to internal stress and leading to the formation of cracks.

Recently in an attempt to create crosslinkable and yet not brittle adhesive bonds, we managed to synthesize allyloxyethyl 2-cyanoacrylate monomer [20].   It combines an unsaturated bond with a long ester radical, giving elasticity to the polymer.   Reaction of the allyl group was confirmed by IR and DSC studies of the formed adhesive bond, when the latter was heated.   Allyloxyethyl 2-cyanoacrylate adhesive bonded steel joints show tensile shear strength of 7.0 MPa, which changes to 14.3 MPa and 6.5 MPa after 24 h ageing of the specimens at 100°C and 150°C.   Interesting results are obtained when allyloxyethyl 2-cyanoacrylate is combined with the widely commercially used ethyl 2-cyanoacrylate (Table 3).

TABLE 3

Contents of allyl-oxyethyl 2-cyano-acrylate in mixture	Lap shear strength (MPa)		
	24 h at 20°C	24 h at 100°C	5 h at 150°C
0	24.2	11.0	5.8
10	21.3	32.6	22.5
30	16.5	32.7	7.5
50	13.6	29.5	8.0
90	7.5	14.8	7.8
100	7.0	14.3	7.5

The tensile shear strength at room temperature steadily increases with the increase of ethyl 2-cyanoacrylate content.   After 24 h treatment at 100°C, the adhesive based on a mixture of the two monomers displays higher strength than the original components.   This synergism is explicitly evident from 1:5 to 1:9 ratios.   Then the lap shear strength of the

adhesive composition is more than 30 MPa, i.e. more than twice the strength
of allyloxyethyl 2-cyanoacrylate and three times the strength of ethyl 2-
cyanoacrylate. After treatment at 150°C, exceptionally high strength is
obtained for the 1:9 composition. The results from impact strength show
better elasticity of the adhesive joint based on the mixed adhesive. These
data indicate that allyloxyethyl 2-cyanoacrylate can be used as a crosslinking component of cyanoacrylate as a means for improving the impact and
heat-resistance of their bonds.

Increasing the viscosity of cyanoacrylate adhesives makes them suitable for porous substrates and gap-containing joints. Polyacrylate, poly-
(methyl methacrylate) [21], poly(alkyl 2-cyanoacrylate) [22], polyacrylic
acid and other polymers have been described as viscosity modifiers for
cyanoacrylates. Thermal treatment or UV irradiation of cyanoacrylate
monomer can also be used for increasing the viscosity [23]. We obtained
good results when 2 to 5% of high molecular weight poly(methyl methacrylate)
is dissolved into the cyanoacrylate. Depending on the quantity of dissolved polymer, the adhesive viscosity can be regulated from 50 cP to
1500 cP (Table 4). The rheological characterization of these compositions
(Fig. 7) shows that they behave like typical pseudo-plastic liquids. At
increased shear rate the dissolved poly(methyl methacrylate) molecules
orient themselves in the direction of the shear force, which leads to a

TABLE 4

Dependence of ethyl 2-cyanoacrylate adhesive viscosity on
quantity and molecular weight of poly(methyl methacrylate)
modifier [24].

Content of poly(methyl methacrylate)	Adhesive viscosity (cP)			
	MW = 670000	MW = 960000	MW = 1300000	MW = 1800000
2 wt%	26	38	48	54
3 wt%	35	63	112	210
5 wt%	119	237	1020	1470
7 wt%	312	562	8650	12200

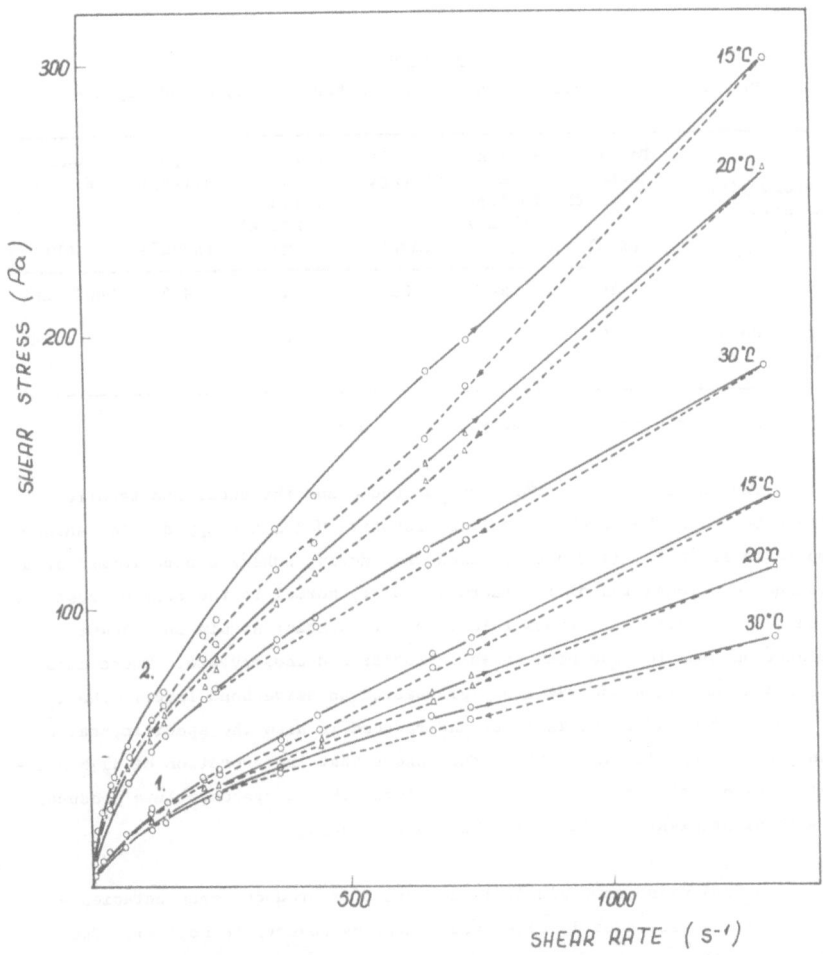

Figure 7.   Dependence of shear stress on shear rate of viscous
             ethyl 2-cyanoacrylate: (1) 100 cP;   (2) 250 cP.

decrease of viscosity value. The physico-mechanical properties of adhesive joints bonded with unmodified and modified ethyl 2-cyanoacrylate (Table 5) shows that the strength remains unchanged. The strain at failure both in shear and tensile mode is increased and the impact strength is improved. The introduction of poly(methyl methacrylate) molecules into the adhesive considerably lowers the value of Young's modulus. On the other hand, the shear modulus is not greatly influenced by the modification.

TABLE 5

Mechanical properties of steel joints bonded with cyanoacrylate

Cyanoacrylate adhesive	Tensile shear strength (MPa)	Strain at failure (shear) (%)	Tensile strength (MPa)	Strain at failure (tension) (%)	Impact strength (kJ/m^2)	Moduli E	G (MPa)
Ethyl	19	24	30	1	4.3	2400	203
Ethyl containing 4% poly(methylmethacrylate)	19	30	30	2	6.2	1300	147

MW of poly(methyl methacrylate) = 1.5×10^6

The values obtained for E and G module and the shear and tensile strain both for the pure and modified adhesive bonds are typical for aniso- tropic materials. It can be assumed that during adhesive bond formation a pronounced orientation of the macromolecules, normal to the bonded substrate, occurs. The data for stress relaxation at constant strain show lower relaxation time for the bond based on modified cyanoacrylate. These data correlate well with the DSC study of removed adhesive bond layer. The T_g of the modified adhesive is lower and reduced is also the specific heat of the glass transition.(Fig. 8). This shows that incorporation of high mole- cular weight poly(methyl methacrylate) into ethyl 2-cyanoacrylate produces a mild plasticization effect on the adhesive bond.

A more pronounced elasticizing effect is produced when butadient- acrylonitrile elastomer is dissolved in the cyanoacrylate monomer. The glass-transition temperature of the resultant bond is further lowered, the Young's modulus is reduced and the time of relaxation decreased. The elas- tomeric component imparts more resilient, toughened structure to the

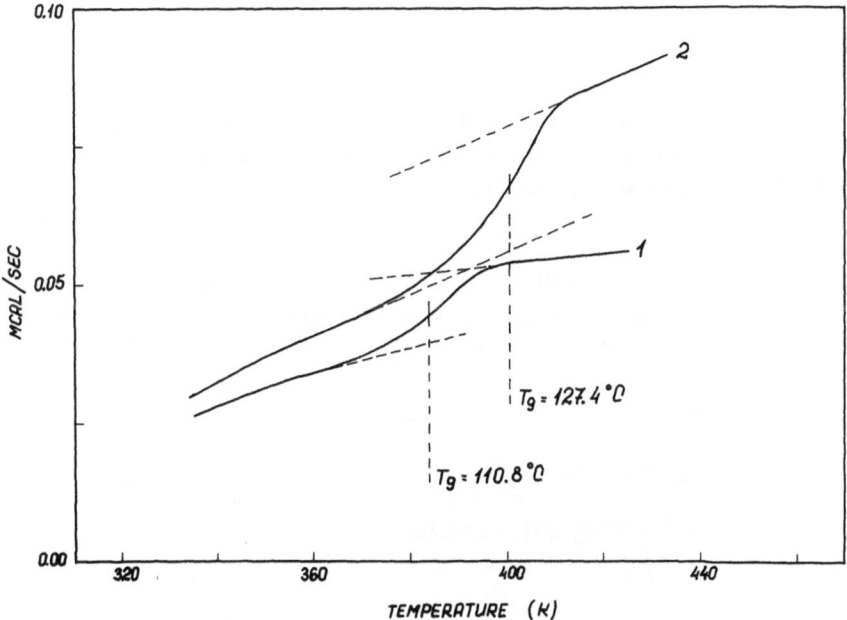

Figure 8.   Glass-transition of ethyl 2-cyanoacrylate
            adhesive bonds:

            (1) pure ethyl 2-cyanoacrylate;   (2) ethyl 2-cyano-
            acrylate containing 4 wt% poly(methyl methacrylate)

adhesive bond.   The peel strength and the durability of the bond to cyclic
deformations is improved.   This adhesive has already found a unique application for express bonding of rubber conveyor belts [25].   To further improve
these elastomer-containing adhesive compositions in respect to practical
preparation (i.e. increasing the solubility of the elastomer into the cyanoacrylate monomer), and on the other hand to enhance compatibility with the
cyanoacrylate polymer, we synthesized two series of copolymers.   They are
based on butadient-acrylonitrile copolymer grafted with ethyl 2-cyanoacry-
late (CABA):

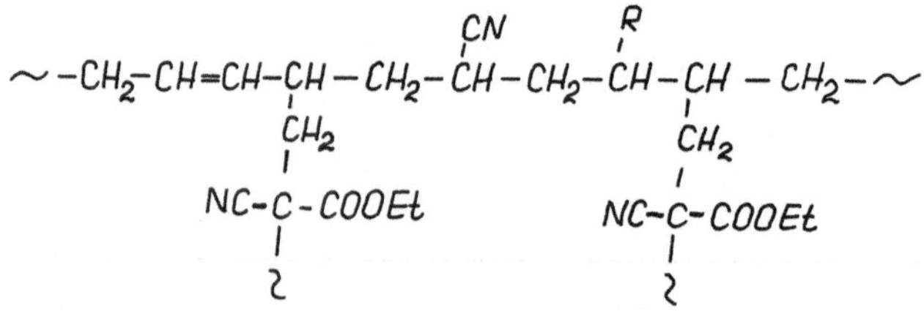

$$\sim -CH_2-CH=CH-CH-CH_2-\overset{\overset{\displaystyle CN}{|}}{CH}-CH_2-\overset{\overset{\displaystyle R}{|}}{CH}-CH-CH_2-\sim$$

$$\underset{\underset{\displaystyle \zeta}{\underset{\displaystyle NC-C-COOEt}{|}}}{CH_2} \qquad \underset{\underset{\displaystyle \zeta}{\underset{\displaystyle NC-C-COOEt}{|}}}{CH_2}$$

and a complex copolymer of ethyl 2-cyanoacrylate, 2-hydroxyethyl methacrylate and glicidyl methacrylate to which ethyl 2-cyanoacrylate and ethyl
acrylate are grafted [26] (CACA).

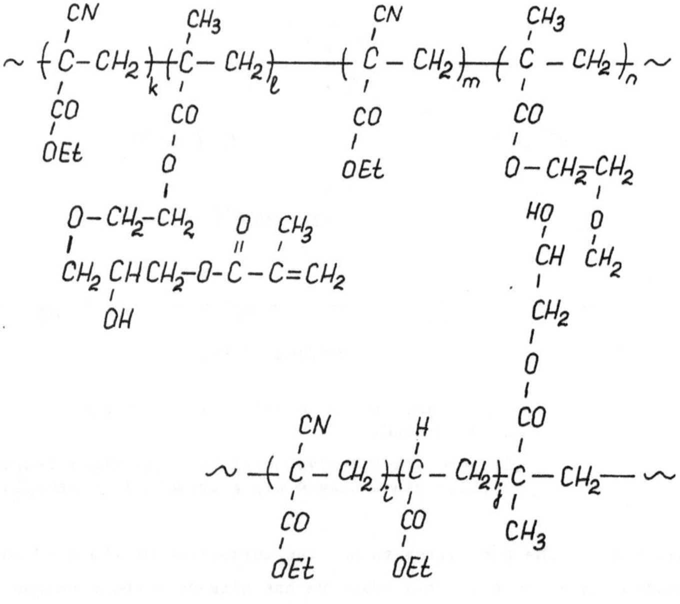

$$\sim \{\overset{\overset{CN}{|}}{\underset{\underset{OEt}{\underset{|}{CO}}}{C}}-CH_2\}_k \{\overset{\overset{CH_3}{|}}{\underset{\underset{O}{\underset{|}{CO}}}{C}}-CH_2\}_\ell \quad \{\overset{\overset{CN}{|}}{\underset{\underset{OEt}{\underset{|}{CO}}}{C}}-CH_2\}_m \{\overset{\overset{CH_3}{|}}{\underset{\underset{O-CH_2-CH_2}{\underset{|}{CO}}}{C}}-CH_2\}_n \sim$$

O-CH₂-CH₂
CH₂ CHCH₂-O-C-C=CH₂
OH

HO   O
CH   CH₂
CH₂
O
CO

$$\sim \{\overset{\overset{CN}{|}}{\underset{\underset{OEt}{\underset{|}{CO}}}{C}}-CH_2\}_i \{\overset{\overset{H}{|}}{\underset{\underset{OEt}{\underset{|}{CO}}}{C}}-CH_2\}_j \overset{}{\underset{\underset{CH_3}{|}}{C}}-CH_2 \sim$$

The improved impact strength, deformability and reduced Young's modulus
combined with the unchanged strength properties of the adhesive bond (Table
6) can be explained by the interaction of the polar groups of the polycyanoacrylate.   Very probably during the adhesive cure, some direct chemical

bonding occurs with the molecules of the polymer modifiers. As a result of the rapid polymerization of their solvent the macromolecules of these elastomers and copolymers are most likely to be finely entangled with the polycyanoacrylate matrix. The morphological structure in the glassy state may be similar to that of an interpenetrating network system, which could account for the considerably reduced time of relaxation and improved creep resistance of the adhesive bond.

TABLE 6

Physico-mechanical properties of elastomer modified ethyl 2-cyanoacrylate bonded joints (steel-steel)

Modifier	Tensile strength (MPa)	Tensile shear strength (MPa)	Impact strength (kJ/m^2)	Strain at failure (shear) (%)	Strain at failure (tension) (%)	Young's modulus (MPa)	Young's modulus (MPa)
None	30.2	19.2	4.3	24.3	1.0	2400	203
Butadiene-acrylonitrile, 2 wt%	28.6	20.5	6.4	25.1	2.0	1680	190
CABA 10 wt%	32.8	18.8	6.9	26.8	3.1	1507	132
CACA 10 wt%	26.4	21.0	8.3	39.0	3.2	1126	121

In an attempt to further diversify the applicability of cyanoacrylate adhesives we tried to obtain electro-conductive compositions. They would have the advantage of rapid catalyst-free cure at ambient temperatures. Carbon and metal powders were tried as fillers. A major obstacle that had to be overcome was the stability of the compositions. An acidic treatment of the fillers did produce compositions which are stable from several days to six months after mixing, depending on the filler [27]. The setting time can likewise vary from 5 min to half an hour. Table 7 shows some typical compositions and their properties. Good results were obtained in terms of stability and aggregation of the compositions, as well as strength of the bond produced when the cyanoacrylate base has viscosity in the range of 50 to 100 cP.

Data for the exponent "n" in the equation giving the dependence between the current intensity and the applied potential $I=cV^n$ (Table 8) shows that only in the case of Ag, Mo, Ni and Fe, Cr, W, n = 1, i.e. there is direct contact between the metal particles in the polymer matrix (ohmic conductivity). In the case of C, Cu, Al, $n \neq 1$, which means that the conductive particles are covered and separated by a polymer film.

TABLE 7

Properties of conductive ethyl-2-cyanoacrylate adhesives

Filler	Size of filler particle (m)	Amount of filler (wt%)	Stability after mixing	Setting time (min)	Volume resistivity of bond (m)	Tensile shear strength (MPa)
Carbon black	0.04	35	5 months	20	2.1×10^{-3}	5.4
Graphite	10	28	4 months	10	1.1×10^{-3}	8.5
Silver	20	70	4 days	6	1.4×10^{-5}	12.0
Nickel	15	70	5 days	9	3.5×10^{-5}	11.3
Molybdenum	10	72	10 days	20	1.0×10^{-5}	10.3

TABLE 8

Conductivity mechanism of ethyl 2-cyanoacrylate adhesive bond

Quantity of filler (wt%)	Value of "n" in $I = cV^n$				
	C Black	Graphite	Ag	Cu	Cr
40	18.3	1.84	1.02	1.92	2.58
50	1.42	1.52	1.00	1.98	1.88
60		1.34	1.00	1.96	1.40
70		1.20	1.00	1.94	1.25
85					1.03

Cyanoacrylates find unique application as surgical adhesives. The higher alkyl homologues (butyl, heptyl, ethoxyethyl.cyanoacrylates) are used as medical adhesives. The reasons are three. Those cyanoacrylates wet and spread on living tissue better than the lower homologues. During polymerization, less heat is evolved, which does not cause necrosis of the tissue. Furthermore, the degradation "in vivo" goes with adequate speed so

that the products of degradation are easily metabolized and removed from
the body [28].   In Bulgaria, after extensive experimentation, butyl 2-
cyanoacrylate has been granted permission to be used clinically. The comprehensive studies have shown that the adhesive strongly bonds live tissues.
It is resorbed in 3 weeks to 1 year depending on the amount applied, without interference with the regenerative process. The wound heals by the same
mechanism as in the case of sutured wound. The adhesive is not toxic,
allergetic or carcinogenic.   As a substance the adhesive is autosterile,
has antibacterial effects and provides for haemo-, bilo-, uro-, and aero-
stasis.   Compared with surgical sutures the use of butyl 2-cyanoacrylate
has the following advantages: The time required for bonding is shorter
than that necessary for suturing; the function of the organ is preserved to
considerably greater extents; the cicatrice formed is tender, linear and
aesthetic; the adhesive can be applied on a large scale in rendering emer-
gency medical aid.   Adhesive bonding can be carried in the course of a
great variety of surgical interventions such as surgery of the parenchymal
organs (liver, kidney, spleen), thoracic surgery (lungs, trachea, bronchi),
urology, neurosurgery, opthalmic surgery, obstetrics and gynaecology etc.
Last year 3500 ampoules of tissue adhesive were consumed in Bulgarian
hospitals and clinics which means that it was used in as many operations or
surgical interventions.

Attempting to diversify the medical cyanoacrylate version we tried
to obtain compositions containing drug substances. Some of the compositions
have already been tested and have given good experimental and clinical
results.   By polymerization at the injured site the cyanoacrylate secures
the drug compound and guarantees its prolonged emission at the desired place.
Some of the compositions made and tested are:

(1)   A suspension of corticosteroid in butyl-2-cyanoacrylate, which
        produces anti-inflammatory and protective action when applied
        on mouth aphtha [29];

(2)   Two component, mixed prior to application composition for treatment
        and post-surgery application of acute cases of periodontosis. The
        active ingredients are anti-inflammatory compounds and tissue rege-
        neration stimulants [30];

(3)   Cyanoacrylate coating containing fluorides [31] produces good
        results when applied on teeth. Fluoride containing sealants
        are also developed [32]. They provoke mineralizing effects
        for a period of over 3 months and help the retentivity of the
        fissures. SEM micrographs show improved structure of the
        surface enamel with mineral depositions. X-ray dispersion
        spectroscopy shows increased content of Ca and P and increased
        Ca/P ratios;

(4)   Last but not least, very promising results were obtained by
        fibroendoscopic treatment of ulcers of the stomach and duodenum.
        The adhesive composition contains drugs that have proved to be
        effective for ulcer therapy. The polymeric film formed over
        the ulcer protects it from the aggressive environment and at the
        same time releases the active drug compounds. The hospitaliza-
        tion period is reduced by one half on the average [33].

REFERENCES

1.   Japan Adhesives Industry Annual Book, Tokyo, 1985, p.57.

2.   Coover, H.W. Jr., Handbook of Adhesives, Skeist, I., Ed., Van
      Nostrand-Reinhold, New York, 1977, p.569.

3.   Guthrie, J., Otterburn, M.S., Rooney, J.M. and Tsang, C.N., J. Appl.
      Polym. Sci., 1985, 30, 2863-2867.

4.   Rooney, J.M., Polym. J., 1981, 13, 10, 975-978.

5.   Coover, H.W., Jr., and Wicker, T.H., US Patent 2765332, 1956.

6.   Braun, B., German Patent 1801711, 1969.

7.   Wolfgant, I., German Patent 2112341, 1972.

8.   Ohashi, K. and Tanaka, T., Japan Patent 89325, 1975.

9.   Kotzev, D.L., Naidenov, A.L. and Kabaivanov, V.S., Compt. rend.
      Acad. Bulg. Sciences, 1986, 39, 8, 53-56.

10.   Coover, H.W., Jr. and McIntyre, J.M., US Patent 3728375, 1973.

11.   O'Sullivan, D.J. and Mebody, D., US Patent 3832334, 1974.

12.   Krall, R., Patent DDR 156365, 1981.

13.   Schoenberg, German Patent 2833842, 1978.

14. Kotzev, D.L., Denchev, Z.Z. and Kabaivanov, V.S., "Adhesive
     Properties of ethyl 2-cyanoacrylate containing small amounts of
     acetic acid as adhesion promoter", to appear in Int. J. Adhesion
     and Adhesives.

15. Bolger, J.C., "Acid base interactions between oxide surfaces and
     polar organic compounds", Adhesion Aspects of Polymeric Coatings,
     Mittal, K.L., Ed., Plenum Press, New York, 1983, pp.3-18.

16. Suetaka, W., "Infrared spectroscopic investigations of polymer
     coatings - metal substrate interactions", Adhesion aspects of
     Polymeric Coatings, K.L. Mittal, Ed., Plenum Press, New York, 1983.

17. Kotzev, D.L., Novakov, P.C. and Kabaivanov, V.S., Angew. Makromol.
     Chemie, 1980, 92, 1421, 41-52.

18. Kotzev, D.L., Ward, T., Kabaivanov, V. and McGrath, J., Polym.
     Prepr. Am. Chem. Soc., Div. Polym. Chem., 1980, 21, 2, 158-159.

19. Kotzev, D.L., Ward, T.C. and Dwight, D.W., J. Appl. Polym. Sci.,
     1981, 26, 1941-1949.

20. Denchev, Z., Kotzev, D. and Kabaivanov, V., Bulgarian Patent 76796,
     1986.

21. Kamlander, L., French Patent 2121597, 1972.

22. Coover, H.W., Jr. et al., US Patent 2794788, 1957.

23. Siro, I. and Narisava, H., Japan.Patent 37277, 1971.

24. Kotzev, D.L. and Dicheva, L.B., First National Conference on
     Chemistry, Sofia, 1985, p.415.

25. Kabaivanov, V., Kotzev, D. et al., Bulgarian Patent 29489, 1979.

26. Kabaivanov, V., Dikov, V., Petrov, C. and Kotzev, D., Bulgarian
     Patent 76320, 1986.

27. Chorbadjiev, K., Kotzev, D. and Kabaivanov, V., Bulgarian Patent
     73341, 1986.

28. Leonard, F., Kollins, J. and Porter, H., J. Appl. Polym. Sci., 1966,
     10, 11, 1617-1623.

29. Pasheva, M., Atanasova, E. and Zaprianova, M., First National
     Symposium on Application of Cyanoacrylate Adhesives in Medicine
     and Technics, Sofia, 1975, p.171.

30. Kabaivanov, V., Kotzev, D. et al., Bulgarian Patent 35202, 1983.

31. Mateeva, C., Kabaivanov, V. et al., Bulgarian Patent 76708, 1986.

32. Mateeva, C., Kabaivanov, V. et al., Bulgarian Patent 76707, 1986.

33. Akimova, A., Private Communication.

POLYMERISATION AND SURFACE INTERACTION
OF A SILANE COUPLING AGENT

P J Grey
BTR Industries Limited
Central Development
Burton-on-Trent
Staffordshire DE13 0SN
UK

INTRODUCTION

The growth of the glass-reinforced plastics industry over the last thirty years might well have been far less spectacular had not organofunctional silane coupling agents been developed as a means of prolonging the useful life of composites when exposed to moisture.

As with many other developments, the technology of silane coupling agents preceded a scientific understanding of their true mode of action. Only recently, with the advent of powerful surface analysis techniques, has the structure of the interface between the glass reinforcement and resin matrix begun to be more clearly understood. A number of excellent accounts of the current state of knowledge have appeared in the literature, included among which are those of Kinloch [1], Comyn [2], Plueddemann [3] and [4] and Kardos [5].

The mechanism by which interfacial adhesion is improved is thought to be due, at least in part, to the formation of primary chemical bonds between the coupling agent and the glass on the one hand and the resin phase on the other. Early studies in this area were undertaken by Sterman and Marsden [6], Evans and White [7] and more recently by Koenig and Shih [8].

Other workers hold the view that the silane affects the mechanical properties of the resin in the immediate vicinity of the interface. Kumins and Roteman [9] and Kwei [10] have suggested the formation of a restrained layer. Erickson and Volpe [11] postulated that preferential adsorption occurred at the interface. The possible influence which surface wettability might have was highlighted by Zisman [12] and his lead was followed by Bascom [13] and Lee [14]. Additionally Plueddemann [15] has proposed a reversible hydrolysis mechanism in which breaking and reforming of stressed bonds between the coupling agent and glass occurs in the presence of water, allowing stress relaxation to take place without loss of adhesion.

In practice most silanes are applied from mildly acidic, dilute aqueous solutions. However, because alkoxysilanes hydrolyse and undergo condensation, their composition at moment of application, and the effect which that might have on the hydrolytic stability of the resulting bond, is only imperfectly understood at present.

For example, Schrader [16] showed that more than 90% of a multilayer film of radioisotope-labelled aminosilane could be removed from a treated glass surface by a combination of cold and hot water extractions. More recently Ishida and Koenig [17] have examined by means of Fourier transform infrared spectroscopy [FTIR] the hydrolytic resistance of coupling agents applied to E-glass fibres from aqueous solution. The findings were similar to those previously reported by Schrader, a very small percentage of the deposited silane remaining on the glass after exposure to water at 80°C.

The acid-base effects of various inorganic oxide fillers have been studied by Ishida and Miller [18] using gel permeation chromatography [GPC] and diffuse reflectance infrared Fourier transform [DRIFT] spectroscopy. A hydrolysate of γ-methacryloxypropyltrimethoxysilane [γ-MAMS] was adsorbed onto various particulate minerals where it was found to form two well defined layers, the relative proportions of which were influenced by the pH of the surface, the surface chemical functionality and adsorbate loading. The physisorbed silane was removed from the surface by means of a gentle wash with tetrahydrofuran [THF] solvent and subjected to GPC analysis. It was found to consist of unhydroylsed or partially hydrolysed monomer and polysiloxane oligomers and higher polymers with molecular weights up to several thousands. The particulate minerals which gave a near-neutral pH when formed into an aqueous slurry [eg TiO_2 and amorphous silica] resulted in almost complete chemisorption of the silane at a 1% loading. The THF extract yielded only a small amount of high molecular weight polysiloxane upon GPC analysis, the mild extraction conditions possibly militating against the removal of any chemisorbed species from the substrate surface. A similar effect was observed with the 'acidic' oxides Al_2O_3, SnO_2 and Fe_2O_3.

The adsorption onto mica of hydrolysates and condensates of γ-MAMS has been examined by Savard and co-workers [20] using ^{1}H-NMR. Whereas hydrolysis was shown to occur slowly, condensation took place rapidly under similar conditions. An induction period preceded adsorption, but if the induction period was too long [eg 1.5 hrs] adsorption was prevented. From these findings it was concluded that long hydrolysis times allowed polycondensation of the silanetriol to either the cyclic trimer or tetramer, which was then ineffective as a coupling agent. Unfortunately ^{1}H-NMR was unable to distinguish between the cyclic trimer and cyclic tetramer.

The work now to be described examines by means of GPC and FTIR the course taken by the isothermal hydrolysis and poly-condensation of γ-glycidoxypropyltrimethoxysilane [γ-GPMS] at pH 4 and $25^{\circ}C$, and the interaction that occurs between THF solutions of the polymerisate and silica gel at various stages during the polymerisation process. Selective adsorption is shown to occur.

MATERIALS AND METHODS

A Fourier transform infrared spectrophotometer with data handling facility [Nicolet Instruments 5-MX optical bench and 1200S data station] was used at a resolution of 4 cm^{-1} with co-addition of 50 scans. Liquid samples in the form of thin films between KBr discs were used, the film thickness being controlled by a 0.012 mm tin spacer.

Analytical GPC was performed at 20-22$^{\circ}C$ using a Waters chromatograph equipped with a refractive index detector and two 600 x 7.7 mm columns packed with 5 μm particle size PL-gel [Polymer Laboratories Ltd] of 100 and 500 Å pore type respectively. The analytical columns were protected by a 50 x 7.7 mm guard column having a PL-gel packing of 100 Å pore size. The mobile phase was THF [Rathburn Chemicals Ltd] and the elution rate 1 ml/min. The columns were calibrated using 0.1% solutions of polystyrene standards [Polymer Laboratories Ltd] having molecular weights [$\overline{M}_n$] of 495, 710 and 900 and a polydispersity less than 1.10. The molecular weight of each oligomeric silane peak reported below was measured relative to its polystyrene equivalent.

The coupling agent [γ-GPMS] was generously supplied by Union Carbide [Silane A-187] and used as received. Silica gel 60H was purchased from BDH Chemicals Ltd [Code Merck 11695] and had a surface area of 365 m^2/g by means of nitrogen adsorption.

Hydrolysis and polycondensation of the silane was carried out at 25$^{\circ}C$ using 3 moles water per mole silane, the water having previously been adjusted to pH 4 with glacial acetic acid. In each case the mass of reactants was 11.80 g γ-GPMS and 2.70 g water [pH 4]. The mixture at ambient was agitated vigorously on a laboratory shaker until a clear solution formed, which normally took 30 ± 5 minutes, at which point the reaction vessel was transferred to a thermostatically controlled water bath at 25 ± 0.1°C. At given intervals 0.5 g of the polymerisate was withdrawn and made into a 1% ^{W}/w solution in THF. This solution was stable for at least 24 hrs. An injection volume of 200 μl was used for GPC analysis, and the composition of the polymerisate was studied after 2, 23, 47, and 71 hrs reaction time.

Adsorption onto silica gel was carried out at 20-22°C using 0.5 g silica gel contained in a 10 ml Pyrex volumetric flask to which sufficient of the 1% ^{W}/w polymerisate solution in THF was added to bring the total volume to 10 ml. The stoppered flask was shaken vigorously for 1 hour, after which the solution was separated for 5 mins at 4,400 rpm on a laboratory centrifuge. The supernatant liquid was clarified using a Millex-SR 0.5 μm disposable PTFE membrane filter [Millipore SA], immediately following which 200 μl was subjected to GPC analysis. A control without silica gel was run at the same time. The peak heights of the monomer and oligomer species were recorded for the polymerisate that had been in contact with silica gel, and the differences in peak heights were calculated relative to the control.

RESULTS AND DISCUSSION

A. Polymerisation. GPC data. Figure 1 shows the general form taken by the gel permeation chromatograms for the γ-GPMS monomer and corresponding polymerisates after 2, 23 and 71 hrs reaction time. Disappearance of the hydrolysed monomer [Peak 1] was clearly evident, and it was interesting to see how quickly Peak 3 overtook Peak 2. The latter became a shoulder on the side of Peak 3 and gradually reduced in intensity as the reaction proceeded through to 71 hrs. Peak retention times and polystyrene equivalent molecular weights for each of the six peaks are given in Table I.

The oligomers corresponding to Peaks 3-6 inclusive could not be assigned definitive structures but, as will be seen later, strong evidence suggested that Peak 3 was due to the cyclic tetramer. Formation of the higher oligomers [Peaks 4-6 inclusive] was perceived to occur by the addition of 2- or 4-mer units. Tentative assignments for the oligomer peaks are given in Table II.

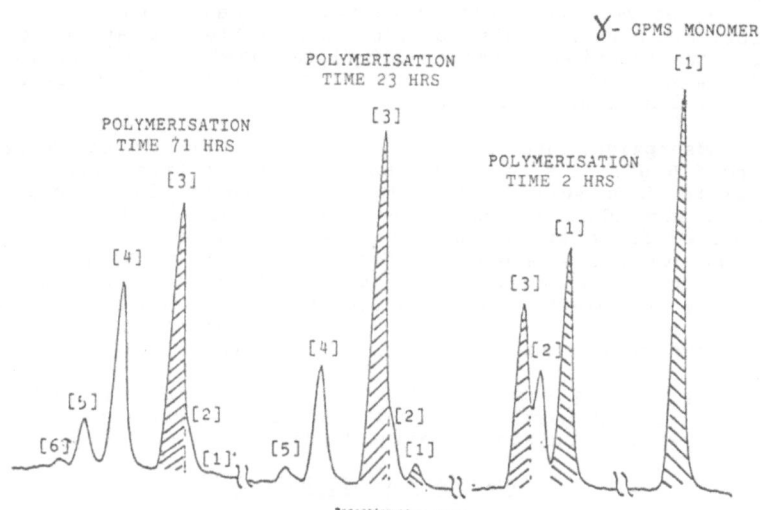

Figure 1.   Gel permeation chromatograms of
γ - GPMS polymerisates

TABLE I

Polystyrene equivalent molecular weights of γ-GPMS monomer
and polymerisates

Peak Number	1	2	3	4	5	6
Peak retention time [mins]	35.00	34.13	33.53	31.73	30.73	29.89
Polystyrene equiv mol wt	119	163	215	393	567	777

TABLE II

Peak assignments for the polymerisate of γ-GPMS

Peak Number	Assignment	Nominal Mol Wt
1	Hydrolysed monomer	208
2	Dimer	398
3	Cyclic tetramer	764
4	Linear hexamer	1130
5	" octamer	1496
6	" decamer	1848

As the polycondensation of γ-GPMS proceeds, the refractive index increases until a limiting value is reached. Allen, Hansrani and Wake [25] noted that the refractive index of γ-GPMS increased from 1.440 to 1.585 when it was subjected to acid-catalysed hydrolysis followed by heat treatment at 70°C for 168 hrs.

In the present work the refractive index of the tetrahydrofuran mobile phase was 1.407. Because of the divergence noted above, the apparent sensitivity of the detector would be expected to increase with increasing degree of polymerisation. Hence the composition of the polymerisate could not be quantitatively determined in the absence of a means of calibration.

The acid-catalysed scission-recombination reactions occurring in organopolysiloxanes have been studied by Damm, Golitz and Noll [26] and found to be first-order with respect to the oxonium intermediate. Noll [27] has determined the half-life for acid-catalysed cleavage of four cylic and two linear polydimethylsiloxanes. The data are shown in Table III.

TABLE III

Half-lives of acid-catalysed scission of organopolysiloxanes

Species	Half-life (hr)
Hexamethylclyclotrisiloxane	0.44
Octamethylcyclotetrasiloxane	175
Decamethylcyclopentasiloxane	310
Dodecamethylcyclohexasiloxane	1000
Linear Hexamethyldisiloxane	36
Linear Octamethyltrisiloxane	780

The GPC chromatograms for γ-GPMS were used to calculate kinetic data for the disappearance of the monomer and each of the five subsequent oligomer peaks. An exponential curve-fitting programme was used for the first-order rate equation

$$\ln [a_0-x] = \ln a_0-kt \qquad [1]$$

where 'a_0' is the initial concentration, as given by the peak area, and 'x' is the amount of peak area that has disappeared in time 't'. Regression coefficients were generally better than 0.985. If the disappearance of a substance follows first-order kinetics the time required for one-half of the original amount present to decay is given by:

$$t^1/_2 = \frac{1}{k} \ln 2 \qquad [2]$$

and therefore the half-life is independent of the initial concentration. The rate constants and half-lives for the silane monomer and five oligomer peaks are given in Table IV.

TABLE IV

Kinetic data for γ-GPMS monomer and polymerisates
[3 moles water pH 4 per mole silane at 25°C]

Peak	Retention time (mins)	Rate constant	Half-life (hr)
1	35.00	-0.0767	9
2	34.13	-0.0149	47
3	33.53	-0.0090	77
4	31.73	-0.0023	301
5	30.73	-0.0018	385
6	29.89	-0.0015	462

The half-life of the dimer of hydrolysed γ-GPMS [Peak 2] was seen to be similar to that of hexamethyldisiloxane. If Peak 3 was attributable to the cyclic trimer of γ-GPMS and not, as is being proposed, the cyclic tetramer, the indications from Table III are that its half-life would be considerably shorter than 77 hrs.

FTIR data. Many studies have been carried out on the infrared absorption spectra of linear and cyclic organopolysiloxanes [21], [22], [23], and [24]. All cyclic trimers were found to have a strong absorption at 1010-20 cm^{-1} [9.8 - 9.9 μ] whereas the corresponding cyclic tetramers had a strong absorbance at 1075-92 cm^{-1} [9.15-9.30 μ]. These absorptions are thought to be associated with the stretching vibrations of the Si-O-Si bond.

The linear trimer and tetramer of polydimethylsiloxane have strong absorbances at 1046-76 cm^{-1} [9.29-9.56 μ] and 1038-80 cm^{-1} [9.26-9.63 μ] respectively. FTIR examination of polydimethylsiloxane standards, obtained from Petrarch Systems Inc., gave the absorbance maxima shown in Table V, which were in close agreement with the values quoted in the literature for Si-O-Si stretching vibrations.

TABLE V

Polydimethylsiloxane standards.
FTIR absorbance maxima

Species	Maxima [cm^{-1}]	
Dimer	1060	
Linear trimer	1050	1076
" tetramer	1043	1079
Cyclic trimer	1018	
" tetramer	1074	
" pentamer	1082	

Interpretation of the FTIR spectra of the raw polymerisates of γ-GPMS was complicated by the presence of methyl alcohol formed as a result of hydrolysis of methoxy groups. Methyl alcohol has a major absorbance at 1029 cm^{-1} [9.72 u] and a less intense peak at 1115 cm^{-1} [8.97 μ]. In an effort to overcome this problem each raw polymerisate was heated for 30 minutes at 35°C under reduced pressure in a rotary evaporator to remove as much methyl alcohol as possible. Ideally the polymerisate should first have been stabilised against further condensation by means of trimethylsilylation [28]. The FTIR spectrum of the 24 hr polymerisate, following this treatment, was typical in that it showed only a slight inflection in the region of 1018 cm^{-1} which was probably due to traces of residual methyl alcohol [see Appendix 1]. Separately it was shown that as little as 10% hexamethylcyclotrisiloxane could easily be detected in the presence of octamethylcyclotetrasiloxane. This evidence, taken together with the kinetic data reported above, suggested that the polymerisate did not contain a cyclic trimer.

An indication of the rate and extent of elimination of methoxy groups during hydrolysis and polymerisation of γ-GPMS was obtained by taking the ratio of the 1192-1200 cm^{-1} Si-O-CH$_3$ peak height to the 910 cm^{-1} epoxide peak. From the earlier work of Allen, Hansrani and Wake [25] it was assumed that the epoxide group remained intact throughout polymerisation of the silane, and that the absorptivity of both bands was tolerably constant. In the present study the

methoxy:epoxide peak height ratio for γ-GPMS was 3.50. As the
theoretical ratio is 3, all data were corrected by the factor
3/3.5 = 0.857.

Within 20 minutes of the commencement of hydrolysis the
average number of methoxy groups remaining per mole silane had
fallen to 1.31. Thereafter the elimination followed first-
order kinetics with a half-life of 46.2 hrs. A graphical
representation of the data is shown in Appendix 2. These
findings were comparable with those of Allen, Hansrani and
Wake [25] who reported that incomplete removal of methoxy
groups occurred after acid-catalysed hydrolysis of γ-GPMS.

B.   **Interaction with silica gel.**   The effect of bringing
10% W/w THF solutions of various polymerisates into contact
with silica gel is shown in Table VI.

TABLE VI

Interaction of γ-GPMS polymerisates with silica gel

Polymerisation Time	Peak Height					
	RT 35.00	34.13	33.53 *	31.73	30.73	29.89
2 hrs Control	24.17	9.65	11.60	–	–	–
2 hrs + Silica gel	24.67	9.06	6.79	–	–	–
Difference	+0.50	-0.59	-4.81			
%			-41.5			
23 hrs Control	2.06	–	51.98	14.09	1.56	–
23 hrs + Silica gel	2.17	–	35.49	14.40	2.52	–
Difference	+0.11	–	-16.49	+0.31	+0.96	–
%			-31.7			
47 hrs Control	–	–	38.04	24.75	6.21	1.35
47 hrs + Silica gel	–	–	27.91	23.04	6.32	1.37
Difference			-10.13	-1.71	+0.11	+0.02
%			-26.6%			
71 hrs Control	–	–	24.05	24.55	8.47	2.44
71 hrs + Silica gel	–	–	17.66	24.59	9.53	3.06
Difference			-6.39	+0.04	+1.06	+0.62
%			-26.6%			

Although the solutions that were used covered the range from hydrolysed monomer up to the oligomer peak having a polystyrene equivalent molecular weight of 777 [Peak 6 as described in Section A], the only species which was adsorbed on to silica gel was that having a peak polystyrene equivalent molecular weight of 215 [Peak 3, retention time 33.53 mins]. In order to precisely define the polystyrene equivalent weight of the species designated Peak 3, analytical GPC of replicate polymerisates gave a mean peak molecular weight of 215 [σ=8] and a mean molecular weight range of 61 [σ=8].

The observed preferential adsorption from mixtures having higher oligomers present was unexpected in view of the statistical model for adsorption of polymers put forward by Simha, Frisch and Eirich [29] who proposed that the total number of anchor points should increase with the square root of chain length. Savard and co-workers [20] suggested that the silanetriol of γ-MAMS and the low molecular weight linear oligomers are the only species that are effective as coupling agents. Plueddemann [3] has also proposed that linear polymerisates of hydrolysed organosilanes are more likely to produce better coupling than the corresponding cyclics.

The distribution and configuration of the isomers of phenylmethylcyclosiloxanes have been studied by Beevers and Semlyen [30]. In the case of the cyclic tetramer the weight fraction of each isomer that would result from the random cyclisation of an atactic precursor is shown in Figure 2.

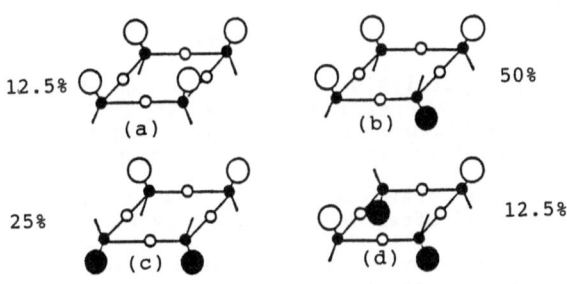

Figure 2.   Structural isomers of $(R^1R^2SiO)_4$ and relative proportions of each

An explanation of the highly specific adsorption onto silica gel of the species designated Peak 3 may be given if the oligomer is the cyclic tetramer. It should be recalled from the FTIR data that the average number of methoxy groups per silicon after 24 hrs polymerisation is approximately 0.75. At that stage the structure of the cyclic tetramer of γ-GPMS would comprise, on average, one silanol and three methoxy groups. Because silanol and silmethoxy groups on the coupling agent are more likely to be attracted to available silanols on the substrate than is the glycidoxy-propyl, organofunctional group, the former are regarded as being the groups which interact with the silica gel surface.

The crown configuration of the cyclic tetramer having one silanol and three methoxy groups on the same side of the plane of the ring [Fig. 2 structure (a)] offers the most sterically favoured arrangement of groups for interaction with the substrate. Adsorption would be expected to occur with the major axis of the cyclic tetramer parallel to the plane of the adsorbate surface. In order to displace a cyclic tetramer molecule a plurality of physisorbed bonds, at least two or three, would need to be severed simultaneously.

Furthermore from Figure 2 it will be seen that structures (a) and (b) account for 62.5% of the isomers present. If a compositional equilibrium existed between the four structural isomers, a depletion occurring in one would be expected to cause the system to move toward restoring that equilibrium, thereby making available more of the species that was being removed from solution as a result of any preferential adsorption process.

CONCLUSIONS

The bulk hydrolysis and condensation of γ-GPMS using 3 moles water [pH 4] per mole silane at 25°C resulted in the formation of a mixture of oligomers, a significant proportion of which appeared to comprise the cyclic tetramer.

The maximum yield of the species designated the cyclic tetramer occurred after 24 hrs polymerisation.

The elimination of methoxy groups was incomplete even after prolonged hydrolysis. FTIR data indicated the number of methoxy groups remaining after 24 hrs to be 0.75 per silicon.

The species designated the cyclic tetramer exhibited a strong, preferential affinity to the surface of silica gel.

These findings suggest it will be possible to enhance and render more consistent the hydrolytic resistance of the technologically important family of glass-reinforced epoxy composites.

ACKNOWLEDGEMENTS

The work reported here is taken mainly from a Ph.D research project carried out at The City University, and in that connection I am especially grateful to Mr K W Allen for his encouragement and guidance. I am also indebted to Mr J B Cameron [Group Technical Director, Permali Limited], Dr W C Wake, and to my colleagues in Central Development, particularly Mrs E Meehan and Dr K Gardner. Finally, I should like to thank the Board of Directors of BTR Industries Limited for allowing publication of this paper.

REFERENCES

1. Kinloch, A.J., J.Mats. Sci., 1980, 15, 2141-66.

2. Comyn, J., In Structural Adhesives, ed A.J. Kinloch, Elsevier Applied Science Publishers, London, 1986, pp. 269-312.

3. Plueddemann, E.P., In Silane Coupling Agents, Plenum Press, New York, 1982.

4. Plueddemann, E.P., In Molecular Characterisation of Composite Interfaces, ed. H. Ishida and G. Kumar, Plenum Press, New York, 1985, pp. 13-23.

5. Kardos, J.L.,   ibid, pp. 1-11.

6. Sterman, S, and Marsden, J.D., Ind. Eng. Chem., 1966, 58, 33.

7. Evans, B. and White, T.E., In Fundamental Aspects of Fibre-Reinforced Plastic Composites ed. R.T. Schwartz and H.S. Schwartz, Interscience Publishers, New York, 1986, pp. 177-197.

8. Koenig, J.L., and Shih, P.T.K., J. Colloid Interf. Sci. 1971, 36,247.

9. Kumins, C.A., and Roteman, J., J. Polym. Sci., 1963, 1A, 527.

10. Kwei, T.K., J. Polym. Sci., 1965, 3A, 3229.

11. Erickson, P.W., and Volpe, A.A., 1963, NOL Tech. Rep. 63-100

12. Zisman, W.A., Ind. Eng. Chem., Prod. Res. Devel., 1969, 8, 98.

13. Bascom, W.D., J. Colloid Interf. Sci, 1968, 27, 789.

14. Lee, L.H.,   ibid   1968, 27, 751.

15. Plueddemann, E.P., Mod. Plast., 1970 47, 92.

16. Schrader, M.E., J. Adhesion, 1970, 2, 129.

17. Ishida, H, and Koenig J.L., J. Polym. Sci., Phys. Ed., 1980, 18, 1931.

18. Ishida, H, and Miller, J.D., Macromolecules, 1984, 17, 1659.

19. Nakatsuka, T., Kawasaki H., Itadani, K., and Yamashita S., J. Appl. Polym. Sci, 1979, 24, 1985.

20. Savard, S., Blanchard, L.P., Leonard J., and Prud'homme, R.E., Composites '83 Symposium, Boucherville, 29 November 1983, Paper 2, National Research Council of Canada.

21. Young, C.W., Servais, P.C., Currie, C.C, and Hunter, M.J., J. Amer. Chem. Soc., 1948, 70, 3758.

22. Smith, A.L., In Analysis of Silicones, John Wiley and Sons, London, 1974.

23. Bellamy, L.J., In The Infrared Spectra of Complex Molecules, Methuen and Co Ltd, 1960.

24. Wright, N, and Hunter, M.J., J. Amer. Chem. Soc, 1947, 69, 803.

25. Allen, K.W., Hansrani, A.K., and Wake, W.C., J. Adhes, 1981, 12, 199.

26. Damm, K., Golitz, D., and Noll, W., Angew. Chem., 1964, 76, 273.

27. Noll, W., In Chemistry and Technology of Silicones, Chapter 5, Academic Press, London, 1968.

28. Sweeley, C.C., Bentley, R., Makita, M., and Wells, W.W., J. Amer. Chem. Soc, 1963, 85, 2497.

29. Simha, R., Frisch, H.L., and Eirich, F.R., J. Phys. Chem, 1953, 57, 584.

30. Beevers, M.S., and Semlyen, J.A., Polymer, 1971, 12, 373.

APPENDIX 1

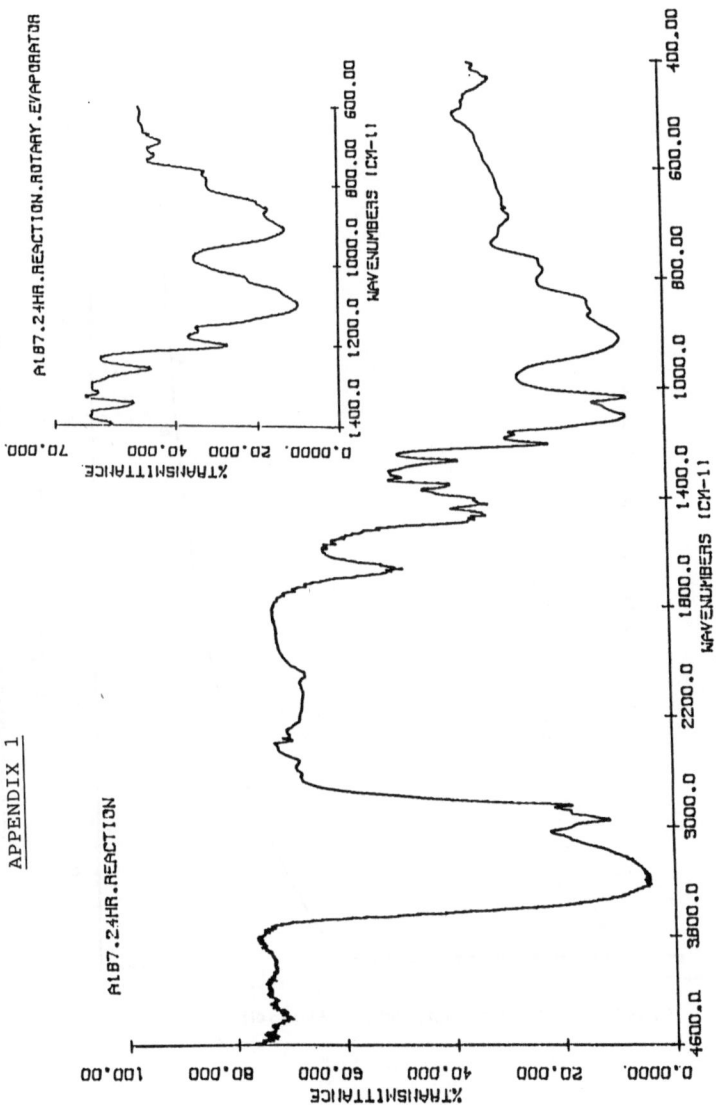

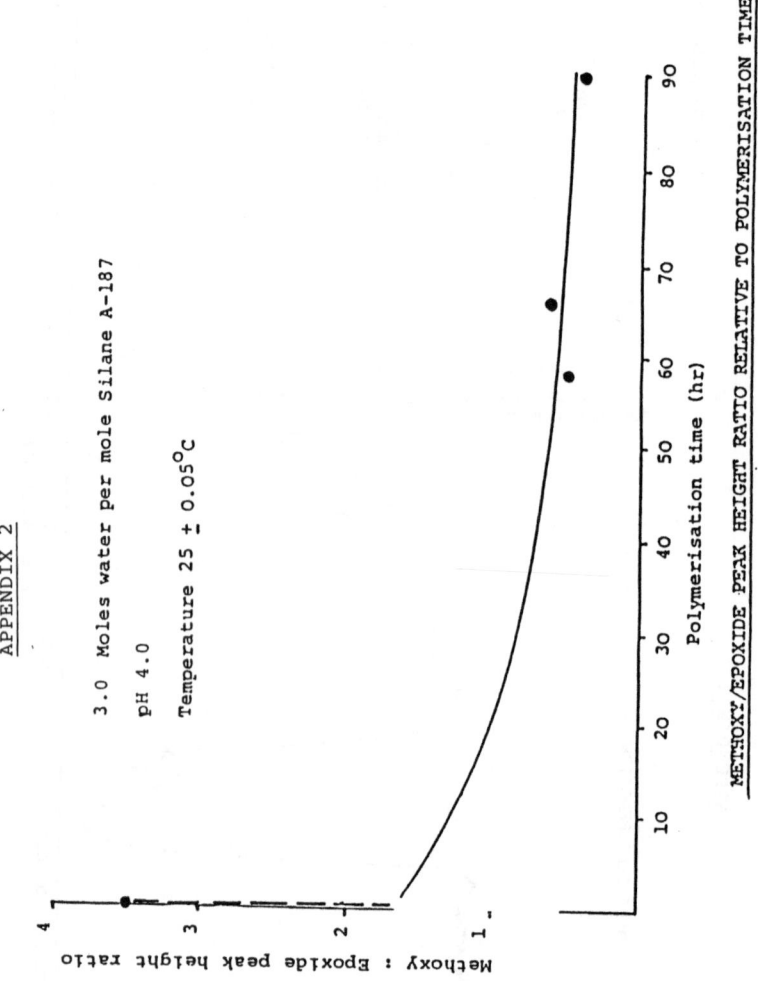

APPENDIX 2

3.0 Moles water per mole Silane A-187

pH 4.0

Temperature 25 ± 0.05°C

Methoxy : Epoxide peak height ratio

Polymerisation time (hr)

METHOXY/EPOXIDE PEAK HEIGHT RATIO RELATIVE TO POLYMERISATION TIME

APPLICATIONS OF TOUGHENED EPOXY AND ACRYLIC ADHESIVES

K W HARRISON
BOSTIK LIMITED
LEICESTER LE4 6BW

One of the significant events of recent years in the adhesives industry has been the development of rubberised or toughened epoxies and acrylics. By this means such desirable properties as tolerance to oily surfaces, the ability to bond dissimilar surfaces over a wide temperature range, high peel strength and impact resistance have been conferred to both systems.

In the case of epoxies, a low molecular weight rubber with functional terminal groups can be pre-reacted with some of the epoxy groups of the epoxide resin. This adduct is still epoxy-functional and so can react with the usual curing agents, amines, amides, anhydrides. But now the properties of the cured polymer have been modified by the inclusion of the rubber, not merely blended in but reacted in.

The rubber is reacted into the acrylic adhesive by different means. It has many reactive sites along the length of its molecule. Each is capable of initiating the growth of an acrylic chain e.g. of methyl methacrylate by a free radical reaction. Such chains terminate when they meet similar chains and so a graft copolymer, a rubberised Perspex, is produced.

In each case the rubber forms a separate internal phase with particles as small as 0.1 microns in diameter, yet the particles are reacted chemically with the external phase and so modify bulk properties. Were they to be discrete they would contribute very little.

Such networks, acrylics or epoxy, offer the ability to dissipate the applied stress over the whole bond area; a crack propagating through the hard phase is stopped on reaching a rubber microsphere.

Epoxy adhesives can be single part, heat cured using a sterically hindered, unreactive curing agent like dicyan-diamide or two part, room temperature curing using amines or amides. None of this changes with the rubber adduct. We shall now follow the fate of this idea through various applications, some where one would not have thought at first sight that the rubber had a role to play.

Bostik E5207, E5210 and E5238 are single part heat cured epoxies whose properties are shown in Table 1. They comprise a series whose only variable is the amount of reacted rubber. Up to about 15% the cured adhesive is brittle and so, although the shear strength at elevated temperature is high, the adhesive is unsafe and E5207 was abandoned. With increased rubber content, the peel strength and impact resistance increase, the shear strength at high temperatures decreases but not markedly.

TABLE 1

Effect of increased rubber content on adhesive properties.

PROPERTY		E.5207	E.5210	E.5238
Added rubber %		10	17	30
Shear strength -30°C (MPa)			37	33
	25°C	31	33	30
	100°C	25	20	
	120°C	20	10.5	6.8
	150°C		4.3	
	200°C		2.6	
	240°C		1.0	0.6
Shear strength after				
1 wk	120°C		33	29
500 hrs	40°C 100% RH		30	20
500 hrs 150°C in HP3 oil			31	27
500 hrs salt spray			30	16
Shear strength, Aluminium, 25°C			33	33
T peel, steel, N/mm		1.4	14	18
Impact resistance, J/cm^2			>4	>10

Substrate = 18SWG car body steel oiled with Croda PQ6 oil, 10mm overlap x 25mm wide, pulled at 20mm/min.   1MPa = 145 psi.

E5238 with the highest rubber content cures to a film so flexible that it can be wrapped around the finger, a peel strength of 18N/mm and an impact resistance such that a supported bond with a bond area of $6cm^2$ can be struck repeatedly with an impact energy of 60N.m without failure (i.e. >10 J cm^{-2}). Yet its hot bond strength is still 7MPa/120°C on oily steel. Reduction of the rubber content to that of E5210 improves the hot bond strength to 11MPa/120°C with a reduction in peel strength to 14N/mm. Both have a shear strength of about 33Mpa on oily steel at room temperature.

BL use E5210, coloured yellow to be visible, to convert the Metro car to the van. A panel has to be bonded over the windows to make the van and this bond had to be authenticated to Customs & Excise that it could not easily be taken out to make the vehicle back into a car. In fact the plate cannot be removed without mutilating the panels so Customs issued the necessary certificate.

Jaguar Cars Ltd tested E5238 very thoroughly and are phasing it into several sites on the vehicle. They have shown E5238 passed the BLS AD01 test for a structural adhesive, it easily passed the corrosion test of 500 hours alternating humidity and salt spray. Used to bond a door clench, it passed the side intrusion test with flying colours. In this test a ram is forced against the door panel and the displacement at which the bond gapes is noted.

Following these tests Jaguar either are using or propose to use E5238 to clench the Daimler Limousine door, to bond the Cabriolet XJS-C B-post and the front head rail (top of the windscreen).

The test results are a good structural argument to use E5238 but Jaguar emphasize the cosmetic improvement they enjoy. The appearance of the highly visible B-post is much improved by the invisibility of bonding compared with welding.

It is now on trial on sports model doors.

An attractive feature for the motor manufacturer is that the same stoving that is needed for the paint can also be used to cure the adhesive.

The ability of epoxies containing rubber to possess a high bond strength at high temperatures is demonstrated by E5314. This is a single part adhesive curing at 120°C. With so low a curing temperature, it is unstable at room temperature and has to be stored frozen. Its special features are a bond strength of 25MPa falling to 18MPa at 135°C when the British Aerospace specification calls for 22MPa falling to 12MPa. The rubber content even manages to give so hard an adhesive some peel strength, 3N/mm which is double the specified value.

E5314 is used on aircraft to bond aluminium to itself and to GRP in sites where the bond can get hot.

TABLE 2
Properties of Epoxy Void Fillers.

PROPERTY		E5260 Actual	BAe Spec	E5358 Actual	BAe Spec	E5441 Actual	RR Spec
Description		Structural		Non Structural		Ultra Light Non Structual	
Colour		Black		Blue		Green	
Density		0.75	0.75	0.5	0.5	0.36	0.38
Compressive Strength MPa	25°C	70	40	14	–	12	7
	135°C	38	25	2.4	–		
Shear strength MPa	25°C	5	3	3.2	2	6.8	3.5
	150°C	6					
Ejection strength KN		4	2.5	2.5	1.5	1.7	1.5
Water absorption %		0.15	2	2	3		
Aircraft Fluids Absorption Average %		0.34	3	0.5	3		
Volatile loss		0.3	1	0.04	1		

Epoxy void fillers are low density materials capable of resisting large compressive loads. Being low density they incur little weight penalty in the structure. Void fillers are used extensively in aircraft honeycomb to reinforce particular areas of it against compressive load and to provide a base for attaching hinges, bolts and whatever else needs to be fitted to the honeycomb.

They are made by the incorporation of hollow glass microspheres of diameter about 50 microns into an epoxy matrix. The matrix must have the ability not to exotherm on curing in bulk because of the large volume of it used.

Bostik has made a range of such products ranging in density from 0.35 to 0.75. Bostik E5260 exceeds the requirements of the British Aerospace specification by half as much again. It has a density of 0.75 and yet can withstand a compressive load of over 70MPa, 10,000 psi, at room temperature and over 35MPa at 135°C. It is still an adhesive giving 13MPa on aluminium falling to 6 MPa at 150°C. This is considered to be structural performance in searching environments like wings, aerilons, tail, undercarriage doors. E5260 is used mainly in military aircraft like the Tornado.

E5358 blue filler is mainly used in civil aircraft. E5358 has British Aerospace specification approval. It has a density of 0.45, is self-extinguishing. Having a compressive strength of only 14MPa, it is used in non-structural applications such as luggage racks, toilet partitions, around hinges, inside internal trim panels and landing light flaps. We have a fast curing modification and even a flexible one which is now on trial at British Aerospace. Its specification calls for a supported cured strip to be depressed at least 12mm over a 66mm length under a force of at least 10N without snapping. We achieved a depression of 19mm with a load of 25N.

Two developments of E5358 are in hand. One, in response to the new stringent fire regulations following the Manchester diaster, is to so choose the flame retardants that not only is the formulation self extinguishing but it has low flame height, low smoke and low toxic gas emission. So far we have a formulation with half the flame height, low smoke emission, acceptable toxic gas emission but which self extinguishes in 18 seconds instead of the target 15 seconds. As the man at British Aerospace said, it is the best he's ever seen yet and we think we know how to knock the remaining 3 seconds off.

The second new development for Bostik, it is well known in the industry, is at Rolls Royce Aero in Derby. Void fillers are used in aero engines as a sacrificial coating on the inside of the outer casing opposite the turbine blades on what is known as the fan track lining. When grit and dust are thrown by the high speed blades against the sides, the void filler is allowed to abrade away so protecting the casing. It has to be replaced from time to time. Blue filler, E5358, is on trial for use directly opposite the blades and E5441, ultra-light filler, colour coded green and with a density of only 0.36 is used next to the blue, further away from the blades to save weight.

The black filler, E5260, is on trial on aero engines in the blocker door on the thrust reverser, core splicing, which is bonding the honeycomb to the panel, smoothing cut edges at the side of honeycombs, on the fan path track and the by-pass duct.

It seems to me that void fillers are capable of wider application. A crash surviving structure is designed to have a sacrificial crumpling zone. The degree to which it crumples could be controlled by reinforcing to a greater or lesser degree with void filler. I made this point in a recent article and I am please to say that two racing car manufacturers have taken up the idea.

In another application, if, for instance, the engineer calculates that the load at a particular site in a box beam structure is too much for the structure to bear then it could be packed with void filler for a few inches on either side of the site to reinforce it against compressive loads. A similar idea has been taken up by a manufacturer of submersible craft.

Another development we have in hand with Rolls Royce Aero is E5464. This is an epoxy adhesive with an anhydride curing agent. Cured at 180°C it shows great strength at high temperature, 17MPa at 200°C, unchanged from its strength at room temperature. It is used in many high temperature sites in the engine for instance the engine suspension mount, on repairs and on honeycombs.

Toughened epoxies even turn up as shimming compounds. The shimming compound E5353 is a mock metal, filled with aluminium powder, designed to bridgegaps between metal spars or a spar and a panel. The special feature of our shimming compound is that after a comparatively long work-life of 80 minutes, the cure then proceeds so rapidly that the polymer can be drilled and machined after 3¼ hours. Here the role of the rubber is to confer some resilience such that when a bolt is tightened through a block of it, there is no tendency to crack or shatter.

E5331 is a two part adhesive specially developed in a joint programme with Floor Plan Electrical Ltd of Blackburn to bond galvanised steel to chipboard in the manufacture of flooring panels. The panels have been tested to a Property Services Agency specification for torsional rigidity, point load bearing and bowing. The adhesive is applied in seven beads side by side from a two component mixing machine equipped with sensors to trigger flow on the arrival and departure of units. The panels are used as raised flooring units in such places as mail order houses, banks, finance houses and computer floors.

Locker Air-Maze of Warrington use E5404 to bond filter end caps. This has been developed through various trials and modifications to become a versatile, all-purpose epoxy adhesive.

It is a two part adhesive capable of curing at room temperature, in which case it is slump-free or, when it is cured on a heated platen at 180°C for $2\frac{1}{2}$ minutes, it self-levels. It is used to bond steel end caps to stainless steel mesh and glass filled polypropylene to Zintec.

In service the bond must resist a variety of environments, some mild as in an engine compartment where it is exposed to a fuel oil mist at 100°F and some hostile as in the filter in gear box oil in hydraulic equipment operating at 120°C, peaking at 150°C. Corena 37 oil is used in compressors. This is a phosphate ester oil and is particularly severe at high temperature because it can so easily react with the adhesive but E5404 resists. It has a bond strength of 17.6MPa which improved on exposure to 19.7MPa.

Reactive acrylic adhesives are as versatile. The difference is they always cure at room temperature. The cure time can be varied over wide limits from about 12 seconds to 45 minutes to allow different assembly times to cope with different processes. They are made by dissolving a rubber in methacrylate monomers, usually methyl methacrylate but higher alkyl methacrylates can be used. Methacrylic acid is added to improve adhesion. The rubber is reactive in that it offers many chain initiation sites for the free radical polymerisation of the methacrylate monomers. These terminate on meeting other methacrylate chains growing out of another rubber molecule to make the tough three dimensional network into which the acrylics cure.

TABLE 3
Rates of Bond Development

ADHESIVE	ACTIVATOR	USED	SHEAR STRENGTH, MPa AFTER GIVEN BOND AGE					
			1 MIN	5 MINS	15 MINS	1HR	1D	3D
M890	A	2 Way	0	9.7	18.4	20.7	22.8	23
M893	A	2 Way	0	0.1	0.6	2.5	16.7	20.7
M896	AR	2 Way	4.8	6.6	10.6	16.6	19.3	21.3
M890	AM	Premix	0	1.4	9.9	21.5	26.7	27.5
M893	AM2	Premix	0	0	1.0	4.6	15.4	17.0
M5435	TM2	Premix	0	4.5	6.8	8.1	17.7	18.8

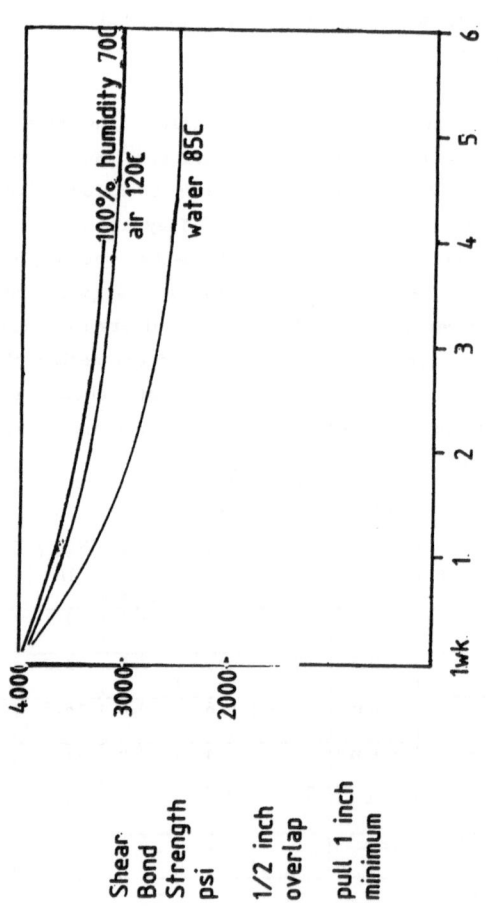

Fig. 1 M890: durability of aluminium bonds

The activator, a reacted amine, which serves to catalyse the cure of the methacrylate solution can be introduced to it in one of two ways. Each component can be applied to one of the surfaces to be bonded. On mating the surfaces the activator diffuses into the adhesive and cures it. However, there is a limit in the gap that can be bridged because cure seals off further diffusion. This limit is about 0.75mm.

The second way, premix, avoids this limit. Activator and adhesive are pumped in ratio through a machine and mixed by a static mixer to give a single curing bead whose time to cure we can control over wide limits. Since cure is assured there is no limit on gap fill. Other advantages are that only one surface is handled and the adhesive can be used as a sealant in that it would fill a crack or make a fillet between two thin sheets at right angles.

In 1977 the GRP minesweeper HMS Brecon used Bostik's reactive acrylic M890 to make thousands of bonds of GRP to itself, aluminium and wood in such applications as conduit, bulkhead partitions, fittings and equipment racks. Since then many of her sister ships have been bonded in the same way. In those ten years of the hurly-burly of a warship at sea, we have never heard a word of complaint.

One of the key properties in its selection was its impact resistance. In Admiralty trials GRP bonds withstood an impact of 45 x gravity, failing, if they ever did, always in the resin, never in the glueline. In trials on steel deck treads bonded to a steel deck, I hammered such a bond 32 times on top and 32 times against the side. The tread was a mangled ruin, but the bond was intact.

We have been engaged for sometime in a joint programme with a well known manufacturer of utility vehicles to develop an adhesive system suitable for the structural bonding of selected sites on a vehicle body.

Some of the early work consisted of bonding a vehicle cab with M895 premix. This was then subjected to a million shakes which it survived without detriment.

In later trials, what can best be described as a cube with two faces removed, was bonded. When compared with a spot welded cube with only one face removed, the bonded structure showed itself to be 12% stiffer in torsional rigidity despite having one less physical constraint.

Other, smaller, structures have shown themselves to be markedly superior in durability and fatigue resistance, than their spot-welded counterparts.

The vehicle company realises, like most of the users of adhesives, that improved strength and durability properties are not the only benefits to be had. Cosmetic appearance, for instance, is greatly improved, secondary seam-sealing becomes unnecessary and sound characteristics are improved due to the inherent damping properties of the adhesive. Furthermore, the manufacturer is keen on the approach of working together with the adhesives industry to achieve the desired result and nowhere is this more evidenced than in assessing the adhesives for their suitability in the manufacturing environment.

Most of the early development work had been done with M890 but due to its relatively low flash point, trials were continued on its low odour derivative, M896. Flash point becomes an important consideration in manufacturing when one considers the amount of electrical machinery present, especially in a body-in-white environment.

Unfortunately, M896 was not happy through the high temperature paint ovens and was quickly replaced by M5435, one of the latest reactive acrylics.

Although it is fair to say that these developments would have happened anyway, they have been speeded by knowing what a potential end-user's needs really are.

We are also engaged in applications trials with Gaydon Technology to assess how these adhesives behave when being dispensed from automatic equipment.

I think the message here is clear, that the adhesives companies must be prepared to work not only with their existing customers but also with their potential ones too. In this way both sides are bound to benefit, the customer in terms of performance and operating costs and the adhesive supplier in terms of having a more saleable product.

British Rail used a premix version of M890 with Activator AM2 to give a work life of 8 minutes and a dejig time of 25 minutes to make the doors for the Class 58 locomotive. This is a locomotive that has 32 doors, 16 on each side. It has so many because it is of modular construction and the doors afford ready access for maintenance and replacement. The doors are made of BSC Galvatite IZ. This has a special zinc surface designed for good adhesion to paint, I find it also accepts adhesives, having obtained 42 MPa on it, that is 6000 psi.

The bonds were first tested in the BR Technical Centre in some of the most thorough testing I have seen. As part of the test programme, they exposed IZ steel bonds at 40°C/95% RH. After 28 weeks exposure the shear strength of M890 premix was unchanged. The competing epoxy and urethane adhesives had halved their shear strengths in 9 and 4 weeks respectively. See Fig 2 page 14.

BR went on to cycle bonds of M890 under load 10^7 times. In their impact testing they found that other adhesives failed at low impact energy by delaminating the zinc layer from the steel because, being of high modulus they concentrated the applied stress at that interface. BR further established that M890, having a lower shear modulus, deformed sufficiently under impact to avoid this concentration at the interface.

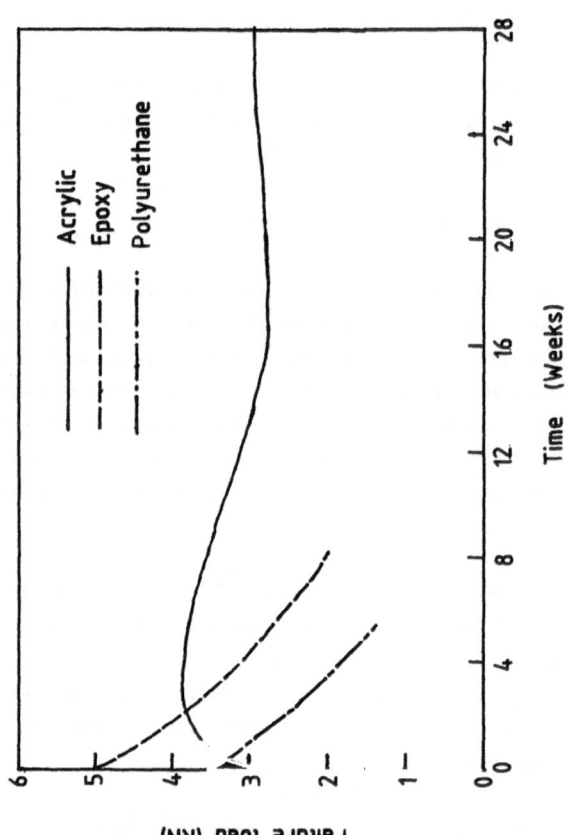

Lap-shear failure loads of 0.8mm IZ steel joists (12.5mm x 25mm) subjected to a 40°C/95% RH durability test

Fig. 2 by courtesy of British Rail Technical Centre

BR told me about an interesting bonus derived from bonding. The fixed connecting plates between the doors could be made easily as a bonded stucture but would have been impossible to weld because there would not have been access.

The Class 58 loco is now undergoing a track trial. It has run around the country for three years now without fault but it is not certain that BR will commercialise this class of locomotive.

The transport industry in general is an important area of growth for our reactive acrylics. Hestair Duple, a leading bus manufacturer, uses 100 cartridges of M893 per bus in the new luxury coach. It is used to bond the panels for the luggage compartment underneath, a weatherstrip of Zintec to stainless steel, GRP to itself for locker doors, GRP to aluminium for the door frame and aluminium to itself for the door hinges, the GRP parcel shelf and the air trunking system. This slow curing version, M893, is used on the luggage doors of the standard range to allow a longer assembly time.

It bonds steel top-hat section stiffeners on to the aluminium skin. Five other Bostik products are used on the bus including a urethane to bond the side panel to the frame so that no fastening point is visible on the outside and another urethane to hold the windows in.

In a TV commercial we made, a BL Mini was suspended from a steel bond of M890 and a circus strong man attacked the bond with a sledgehammer throughout the whole commercial. The bond held. In fact the same bond was used the whole day through all the re-takes, a vivid demonstration of load bearing under impact.

Reactive acrylics have found application in the manufacture of GRP ammunition boxes for parachute drop to troops in the field. Here the slow curing M893 was used to allow all six sides of the box to be coated and then clamped together in one operation. Ferranti and the Army did joint trials in which boxes bonded with the conventional epoxy and with M893 were loaded with 100lbs of sand and dropped from a height of 3 feet onto concrete. The epoxy bonds often failed, if an acrylic bonded box failed, it was always in the GRP, never in the glueline. In an accident a box fell 14 feet onto concrete, the moulding shattered but the glueline was intact.

Several manufacturers make double glazed factory roof-lights of GRP or PVC. They use both an epoxy and an acrylic, both have to be slump-free. The epoxy is cured under infra-red lamps but the acrylic, M895 premix, cures at room temperature. The perimeter of the units can be up to 20 metres long so the adhesive has to have a long work-life to allow application and assembly. It has to be slump-free so that the bead does not flow down from the hills of the corrugations into the valleys. Bonded units have been up in roofs for about 8 years without fault.

Perhaps in a talk about applications I might digress to tell the story of a non-application, a disaster in fact. The moral of the story is to choose the right adhesive for the job. Bostik makes what we call the Reactive Bonding System. That means there are several acrylic adhesive and several activators, any of one can be paired with any of the other. These combinations are designed for different jobs. M896, a low odour, low flammability version of M890 in combination with Activator AR is very rapid curing. Its bond is finger-tight in 25 seconds but it achieves that speed at cost of peel and impact resistance.

Unfortunately British Airways chose this combination for the Boeing Wedge Test. They were severely critical, the results were terrible. In reply we commissioned City University to do the same test on a combination of my choice. The results justified my faith. In the first paper M896/AR gave a fracture energy of 0.1 KJm^{-2} and a crack length of 74mm. In the second piece of work, M890 premix gave a fracture energy of 4.8 KJm^{-2} and a crack length of 28.8mm with zero growth of the crack on immersion in water. My friend at British Airways says a fracture engergy of $2KJm^{-2}$ is good so 4.8 saved the day.

The Tornado fighter plane uses M890 to bond the vortex generators onto its tail. They are little brackets which offer a series of mini-tails to the air stream to increase the grip of the tail on the air and so increase the aircraft's manoeuvrability in a dog-fight. Thus the bond withstands considerable vibration and shear at both very high and very low temperatures. Not one bond has failed in service to date.

And so the story goes on. Next month's applications will be differenct from this month's. New properties will be required to meet new design criteria, but I feel confident that there is sufficient formulation scope in toughened epoxies and acrylics to meet those demands.

Extracts from reference 11 are reprinted from "Structural Adhesives in Engineering" by permission of the Council of the Institute of Mechanical Engineers.

REFERENCES

1. M890 on HMS Brecon.
    Engineering Today - 21.11.77, 13.

2. M890 Reactive Bonding System.
    Plastics and Rubber Weekly - 02.12.77, 14.

3. Harrison, K.W. Rapid Bonding Acrylic Adhesives.
    Fastening - Vol 3, 1979, 25.

4. Bostik Sticks Hope on New Glue.
    European Plastics News - Feb 1980, 55.

5. Mellor S.J. Acrylic Engineering Adhesives.
    Polymer, Paint, Colour Journal - 04.03.81, 10.

6. Harrison K.W. Acrylic Engineering Adhesives.
    Paint Manufacture and Resin News - May/June 1981, 23.

7. Acrylic Adhesives are a Powerful Design Tool.
    Design Products and Applications - May 1982, 11.

8. Seeds A. Future Use of Structural Adhesives in the Construction of
    British Rail Rolling Stock.
    International Journal of Adhesion and Adhesives - Vol 4, No 1,
    Jan 1984, 17.

9. Adhesives in Action.
   International Journal of Adhesion and Adhesives - Vol 5, No 1,
   Jan 1985, 51.

10. Allen K.W., Greenwood L, Armstrong K.B., A Comparison of Different
    Grades of an Acrylic Adhesive for Bonding on Aluminium Alloy.
    International Journal of Adhesion and Adhesives - Vol 5, No 3,
    July 1985, 149.

11. Structural Adhesives in Engineering.
    Institution of Mechanical Engineers Conference, University of
    Bristol - 2-4 July 1986, 179.

12. Adhesives, Sealants and Encapsulants.
    ASE 1986 Conference, Kensington Exhibition Centre, London -
    4-6 November 1986, Day 3, 128.

Slump-free epoxies and acrylic adhesives used to bond
GRP and rigid PVC double-glazed factory roof-lights.

by courtesy of Doleport Ltd., Kenilworth.

The two-part Reactive Acrylic Adhesive, M.890, Activator A,
used to bond loudspeaker magnets.

A Tornado in flight.

by courtesy of British Aerospace PLC., Warton.

The Duple Integral 425 uses several Bostik products in its construction.

by courtesy of Hestair Duple Ltd., Blackpool.

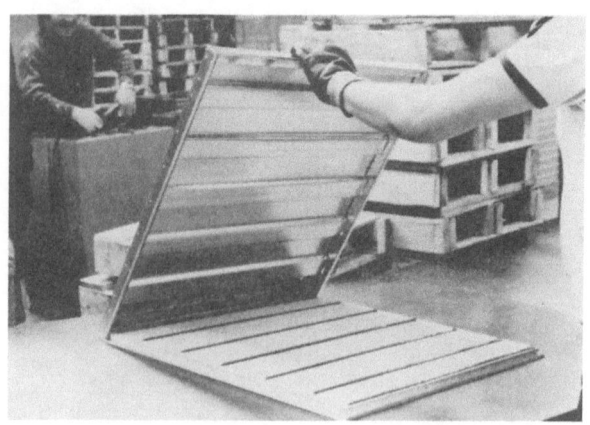

A two-part room temperature curing epoxy adhesive used to bond raised access industrial flooring units.

by courtesy of Floor Plan Electrical Ltd., Blackburn.

STRESS IN BONDED JOINTS

Dr W.A. Lees
Permabond Adhesives Limited

1. INTRODUCTION.

To most, if not almost all, engineers adhesives are
unfamiliar materials - hardly surprising given the lack of
training in the subject. Similarly, very few adhesive
technologists appreciate what an engineer needs to know about
their adhesives in order to be able to predict their
behaviour.

Set against this background the confusion and
misunderstanding seen in both industries seems inevitable.
The confusion is compounded by the fact that it is not only
difficult, and time consuming, to measure an adhesive's
engineering characteristics but these values change in
accordance with the way in which they are measured! An issue
which is taken up later.

It is regrettable - but it must be admitted - that engineers
cannot yet anticipate truly quantitative replies to the
questions they pose. Not even the latest computer based
models of joint behaviour can take into account all
performance variations, particularly those brought about by
impact. Nor is there yet a means of assessing cumulative
damage. What the sympathetic adhesive technologist can offer
is experience, backed by the simple design rules developed
from empirical computerised modelling and the knowledge of
how structures have actually behaved when loaded. For their
part, engineers must adapt their design strategy to a
material which has neither the stiffness nor the cohesive
strength of metal and which, in the last analysis, behaves
best when compressed.

The issues which appear to give engineers the greatest
concern are durability, design and adhesive selection.
Durability is a major subject in its own right. Consequently,
the main purpose of this review is to relate the basic
principles of design to load distribution and the subsequent
selection of adhesives.

2. PROBLEM AREAS.

The demanding conditions faced by stressed structures -
particularly vehicle impact at temperature extremes -
highlight the principal problems which beset adhesives and
which must be faced to some degree whatever the application.
The difficult areas are listed in Table 1 and, taken as a

142

whole, it must be admitted that they constitute a formidable
barrier to the introduction of adhesives in any field of
production.

Table 1 :   Adhesives - Problem areas :-

* Collapse during high speed impact and the
  associated problem of
* Susceptibility to failure when subject to peel
  and cleavage forces.
* Variation in properties with time, temperature
  and humidity.
* Shortage of accurate, useful design data.
* Shortage of usable data relating to
  environmental resistance.
* Total lack of codes of practise and standard
  design solutions.
* Communication between manufacturer and user
  relatively poor compounded by general lack of
  comprehension.

The situation is exacerbated by the practical problems
encountered when a new technology is introduced into an
established industry.

3.   HOW ADHESIVES BEHAVE.

The difficulties listed in Table 1 are related to the basic
characteristics of adhesives and these can be advantageously,
if loosely, reviewed from two points of view -

* The fundamental properties of adhesives.

* The response of adhesives to stress - both
  mechanical and environmental.

Such an examination is both useful and re-assuring for it
places the problems in the correct perspective and indicates
quite clearly what needs to be done in order to accommodate
the less appealing aspects of an adhesives nature.

3.1   The fundamental properties of adhesives.

The types of adhesives likely to be of most use to the
structural and mechanical engineer range from the ductile
polyurethane and acrylic based types through the stiffer
epoxides to the stiff and possibly brittle anaerobics (again
acrylic based) which are so often used in the assembly of
mechanisms.   Even stiffer are the heat resisting adhesives
based on imide and related polymer chemistry, which are
currently used for advanced aerospace structures and which
may well be used in a future generation of engines.

A crude assessment of the quantitative relationship between
them may be obtained from Fig.1 which illustrates the
stress/strain relationship of the two extreme types likely to
be encountered and the properties of two epoxy types
possessing intermediate characteristics.

It is immediately apparent that the Shear Modulus of the
stiffest adhesive is but a fraction of that of any commonly
used engineering metal and also that their cohesive strength
is low. However, when strained to failure many adhesives
display enormous accommodation compared to metal. Thus
adhesives should not be expected to support high loads in
tension (Fig. 2) even though they can resist high shear
loads when the joint overlap is sufficient to allow the load
to be distributed adequately.

This point is considered in detail later; for the moment it
is worth pointing out that asymmetrical tensile loads (Fig.3)
induce peel and cleavage forces which become extreme in the
"T"-peel test (Fig.4) where all the loads are induced and
focused on one end of the joint and where the degree of
strain is so great that not even the most accommodating
adhesive can cope.

The lesson is not to load adhesives in a tensile mode. This
is not so simple, but before considering this and the wider
aspects of design, it is necessary to examine the commonly
used lap joint in detail.

3.1.1 The Lap Joint

The capability of adhesives to carry shear loads has been the
subject of much academic study concerning the relationship of
the material properties of both adhesive and adherend, the
overall geometry of the joint, the detailed geometry of the
joint's edge and the related effect and influence of the
excess adhesive fillet which is almost invariably formed
there.

This work has shown that detailed attention to edge geometry
and fillet shape can bring about major improvements in
performance. But, this level of benefit is probably only
worthy of consideration when the ultimate performance is

required - perhaps in aircraft or missile structures or when
power transmission must take place through the medium of an
adhesive in shear.

Such refinements will be discussed later but, for the moment,
it is more important to come to terms with a few of the
unavoidable facts relating to the basic performance of a lap
joint.

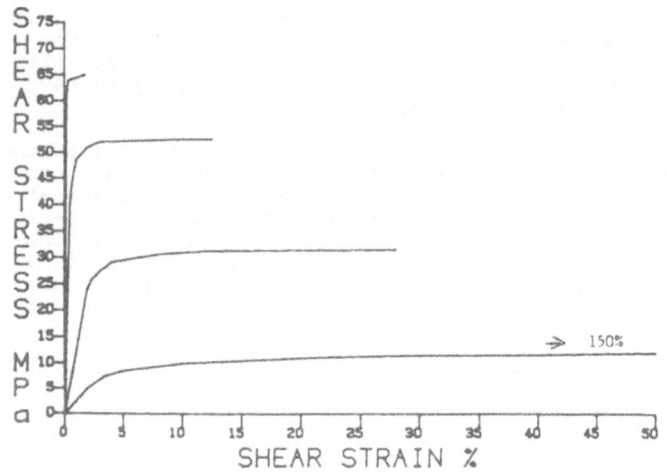

Figure 1.   Typical characteristics of adhesives used for structural
and mechanical engineering assembly.

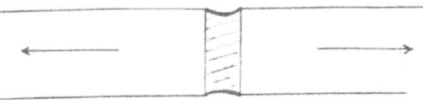

Figure 2.   Tensile load on joint.

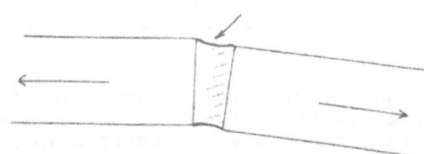

Figure 3.   Offset loading creates
high asymmetrical cleavage forces.

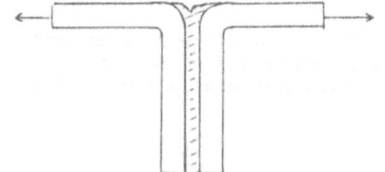

Figure 4. The 'T' peel test induces
irrestiable cleavage forces and
ultimately the adhesive fails
cohesively -but, stiff strong
adhesives will only fail after the
metal (Aluminium) has deformed
plastically.

Perhaps the most important point that can be made is that the lap joint can never be loaded in pure shear - even the strapped double lap (Fig.5), which would appear to allow symmetrical loading, distorts due to the way in which strain is induced.

This means that unless a positive design feature is introduced to <u>prevent</u> cleavage and peel forces developing a lap type joint is always at risk when high loads are likely - by either intent or accident. However, having said this, it is appropriate to point out that the loads seen in most situations are readily borne by the present generation of "Toughened" adhesives and it is only when demands are extreme that there is a need for concern.

The stress distribution of the shear vector of a loaded lap joint is important for determining the overlap necessary and the most appropriate type of adhesive. While these factors are intimately inter-related it is simpler and convenient to consider them separately.

Practical experience has established that there is an upper limit to the load which can be carried by a lap joint no matter to what degree the overlap is extended. It is now possible to simulate this behaviour on a computer using models capable of handling elastic and plastic strain in the adhesive. Curve M of Fig.6 illustrates the point being made and is one of a series of simulations carried out for steel lap joints bonded with a toughened, stiff epoxy adhesive. Taken in conjunction with others in the sequence the overall picture questions the validity of the almost obsessive attempts made, by all but a few enlightened workers ,to determine the ultimate strength of a joint and to use the data so obtained for design purposes.

It is the ability of a joint to resist creep and fatigue which is crucial not its ability to sustain a high load - once!

The calculated relationship of creep, fatigue and ultimate strength to overlap length is illustrated for a model joint in Figure 6 and represented by curves C, E and M. The performance of a complete structure is reviewed in Section 4 -Design.

The three curves M, E and C obtained for the simulated joint show that -

* little benefit, in terms of maximising load bearing capacity, can be obtained by extending the overlap beyond 35 mm (Curve M).

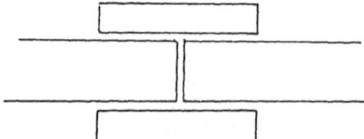

Figure 5. Even this joint design, the strapped double lap, induces asymetric loads in the adhesive as a consequence of the manner in which the metal distorts.

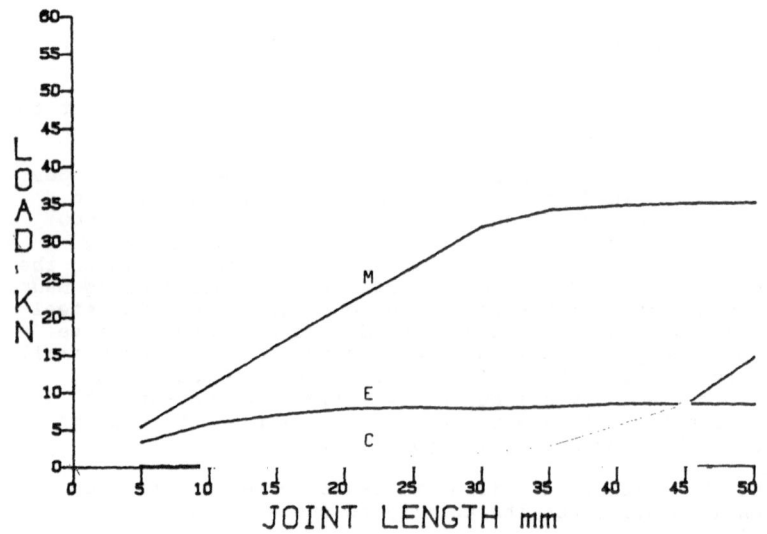

Figure 6. Variation in the load bearing capacity of a model lap joint with overlap length.

Adhesive :   Stiff epoxide, shear modulus 1.5 GPa, elastic limit
              30 MPa, Ultimate strength 45 MPa, 0.8 mm thick.
Adherend :   Mild steel, 2.0 mm thick, width 25 mm.
Line M   :   Maximum load which can be borne by joint before
              rupture for any given length.
Line E   :   Maximum load which can be borne by joint without
              exceeding elastic limit of adhesive at joint's edge.
Line C   :   Maximum load which can be borne by joint while
              ensuring that at some point in the joint's centre the
              load does not exceed 2 MPa.

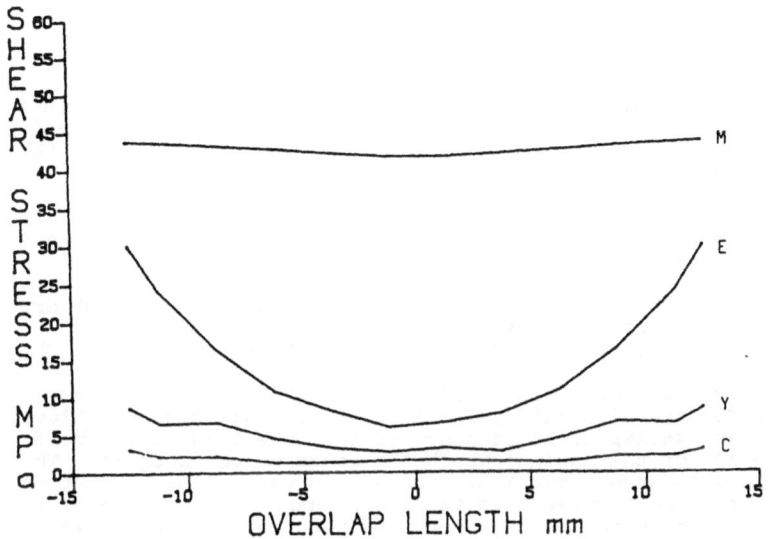

Figure 7. Variation in stress distribution along the length of a model lap joint, length 25 mm, for four given loads.

Adherend :   Mild steel, 2.0 mm thick, width 25 mm.
Adhesive :   Stiff epoxide, shear modulus 1.5 GPa, Elastic limit 30 MPa, Ultimate strength 45 MPa, 0.8 mm thick.
Line M   :   The maximum load (26.7 kN) that may be borne by this joint.
Line E   :   The maximum load (7.9 kN) which can be borne by joint without exceeding the elastic limit of the adhesive.
Line C   :   The maximum load (1.3 kN) which can be borne by joint while ensuring that at some point in the joint's centre the load does not exceed 2 MPa.
Line Y   :   Stress distribution for a load of 2.67 kN (10% of M).

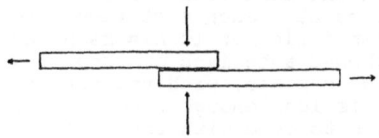

Figure 8. Model lap joint subject to both shear and compressive loads.

* If the load per unit area is to remain within the elastic limit of the adhesive used then it is not really effective to extend the overlap beyond approximately 20 mm (Curve E).

* 45 mm of overlap is required to maximise the load (actually only a marginal benefit relative to a 20 mm joint) which can be carried within the elastic limit of the adhesive while ensuring, by appropriate stress distribution, that the adhesive will not display creep.

Thus, for the simple case of the pure shear vector there are already complications. Clearly, to avoid fatigue the elastic limit must not be exceeded and therefore the ultimate capacity of the joint to carry load is not particularly significant; except in impact overload when other issues are more important. To maximise the working load the best compromise between the three curves must be found.

Further refinement and examination of the calculation shows that for the simulated joint in question this can be achieved with an overlap of 25 mm provided that the load is not more than 10% of the maximum which the adhesive can support. This follows because Curve Y of Figure 7 is very close to that of the "no creep" Curve C in the significant central area of the joint. It is commonly, and probably correctly, believed that provided some part of a joint is essentially unstressed then the joint will not creep. What has not been assessed, as far as is known, is to what degree adhesives may be stressed before they creep significantly, throwing load onto hitherto unloaded areas, and what relationship this load may have to the elastic limit and the modulus. And, in addition, how time and temperature are involved. However, the present empirical calculation indicates that at the point of rupture the maximum load that can be carried by this simulated joint is about 27 kN. Thus, the working load ought not to exceed approximately 2.5 kN. at which level neither fatigue nor creep should occur.

While this load limit may seem modest it is more than adequate to cope with most structural assemblies. For example, a simple comparison of the total structural weight of a vehicle with the bond area available will readily demonstrate this point. Such a statement could be said to be unjustifiably simplistic but it has been made to emphasise that modern toughened adhesives can cope readily with shear loads in the joint's plane, with the not unreasonable proviso that the overlap is long enough. However, they cannot and never will be able to cope with the peel and cleavage forces induced by plastic deformation of a sheet metal adherend. At the moment, it is not possible to calculate readily the relationship between asymmetric tensile loading (peel and cleavage) with creep and fatigue. While strenuous efforts are being made to remedy this situation one cannot help but raise two questions -

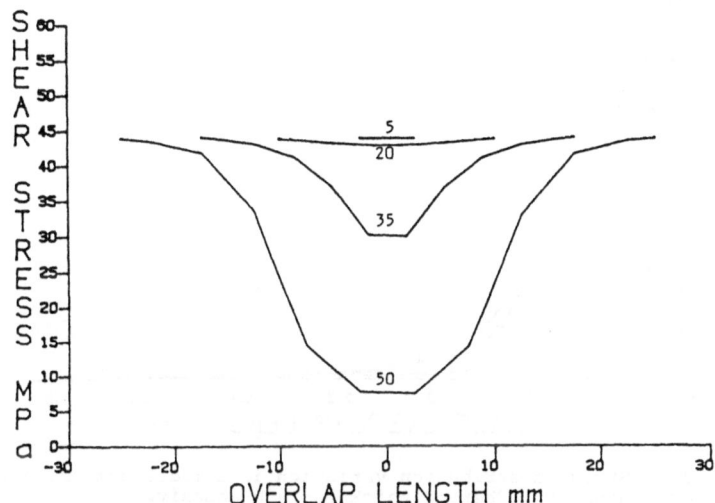

Figure 9. Variation in stress distribution with load and overlap length for an otherwise identical series of four lap joints.

Adherend	:	Mild steel, 2.0 mm thick, width 25 mm.
Adhesives	:	Stiff epoxide, shear modulus 1.5 GPa, Elastic limit 30 MPa, Ultimate strength 45 MPa, 0.8 mm thick.
Line 5	:	5 mm overlap, maximum load 5.5 kN.
Line 20	:	20 mm overlap, maximum load 21.6 kN.
Line 35	:	35 mm overlap, maximum load 34.1 kN.
Line 50	:	50 mm overlap, maximum load 35.1 kN.

"Is it necessary? " and "Does such a lack of knowledge prevent progress?".

While knowledge would give better understanding the practical answer to both questions is "No" for reality demands a simpler approach. All tensile loads should be either suppressed by mechanical means -welds, rivets etc. or should be totally avoided by ensuring that the adhesive within the joint experiences only combinations of the shear and compression loads seen in an idealised form in Fig.8. This is achieved more readily than might be supposed - an issue discussed in some detail in section 4 - Design.

Reference has already been made to the falling efficiency of a joint to carry load when the overlap length is increased - see Curve M of Fig.6. An effect which is caused by the inability of the adhesive/adherend combination to transmit stress to the central areas of the joint. This is exemplified in detail in Fig.9.

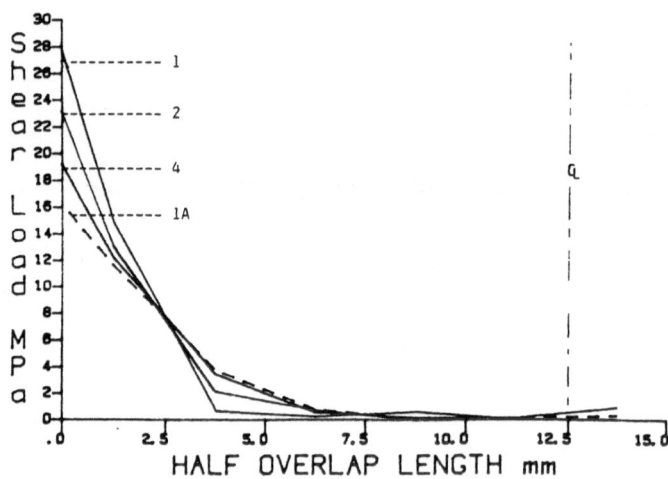

Figure 10. Stress distribution variation in a model composite lap joint with change in adherend thickness and adhesive.

Adherend	:	SMC [Polyester] 1,2 and 4 mm thick sheet.
Adhesive	:	Either stiff polyester or ductile acrylic.
Joint	:	25 x 25 mm, variable adherend thickness.
Line 1	:	Polyester bonded, 1 mm thick SMC
Line 2	:	Polyester bonded, 2 mm thick SMC
Line 4	:	Polyester bonded, 4 mm thick SMC
Line 1A	:	Acrylic bonded, 1 mm thick SMC.

The importance of selecting the correct adhesive is demonstrated by a further examination of stress distribution in a lap joint. For example, in Fig 10 the load at the extreme edge of the joint -bonded SMC - is shown to be reduced by the use of a ductile, toughened acrylic adhesive instead of the stiff polyester based composition. This is clearly beneficial and it is the reason why, whenever possible, ductile adhesives should be used to maximise load distribution.

Furthermore, it can be seen that despite the benefit of thick adherends the polyester induces a greater degree of stress than does an acrylic - even when the latter is used on much thinner, and therefore less stiff, adherends. High edge stresses lead to poor performance. An appreciation of why this is likely to occur and how it may be avoided by the selection of the most appropriate adhesive is crucial to good design.

151

Section 3.1.2   Adhesive Properties -Variations induced by
                  stress and other factors.

Bonded structures and mechanisms which are used in benign
environments are not likely to see the levels of stress
experienced by dynamically stressed vehicles which may be
used anywhere in the world. For this reason, vehicle
stresses are a good yardstick to assess how well adhesives
cope with, or fail to accommodate, the engineer's
requirements.

Thermal Effects.

All adhesives display a major change in their modulus,
elastic and plastic characteristics when the temperature
changes. This is shown for a typical heat cured, single
part, toughened epoxide in Table 2. The change is
particularly marked when the glass transition temperature
[Tg] is passed (for these materials the Tg lies in the region
of 120 -150 C). In view of what has already been said, it is
not surprising that such changes affect the stress
distribution pattern. This is particularly important when
adhesives are used to transmit power directly through shear
loading. It is not so important in structures where, with
the exception of safety critical areas, neither the stiffness
nor the strength of the adhesive will have an unacceptable
effect on the overall performance of the bonded structure.
This point is indirectly demonstrated by the bonded beam
illustrated in Fig.11. where the flexural and torsional
stiffness variations (Table 3) brought about by changing from
a relatively stiff adhesive to a more ductile one may be
compared.

The Influence of Moisture.

Superimposed upon the thermal effects will be the reduction
in stiffness brought about by water absorption. Very

Table 2 : Characteristics* of a typical, heat-cured, toughened
epoxide and their temperature variations.

Property	Temperature °C				
	-40	20	50	100	150
Shear Modulus GPa	2.88	2.43	1.71	1.44	0.045
Elastic Limit MPa	35.1	21.6	13.5	8.1	1.62
Strain at Failure %	18.0	16.2	21.2	26.1	45.0

* Determined using Thick Adherend Shear Test.

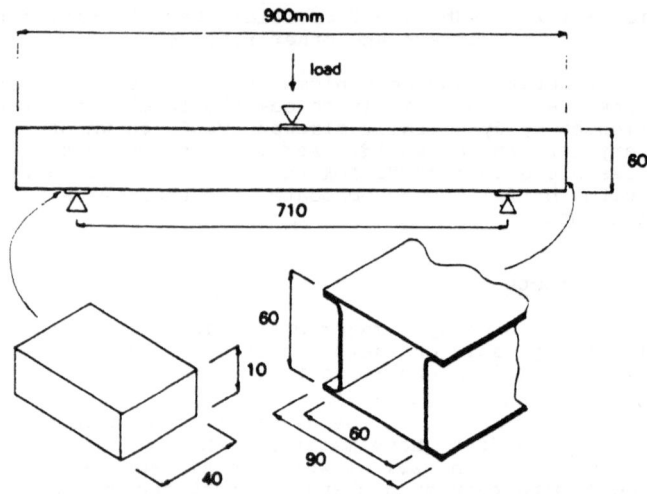

Figure 11. Detail of the rivetted, welded and bonded steel (0.9 mm) beams assessed for both flexural and torsional stiffness and ultimate flexural load bearing capacity.

Table 3 : Properties of the Thin Sheet Fabricated Box Section Beam of Figure 12.

Beam Material	Joining Method	Flexural Stiffness kN.mm^{-1}	Torsional Stiffness kN.mm^{-1} rad	Ult. Load In Flexure kN
0.9 mm Mild Steel	2 part cold cured Epoxy adhesive Shear Modulus - 0.4 GPa	2.9	11.6	7.7
Beam Weight 2.3 kg	Toughened Acrylic Adhesive Shear Modulus - 0.2 GPa	2.85	11.0	5.0

NOTE: Spot-weld and rivet pitch = 25 mm.

considerable amounts may be absorbed by an adhesive in a warm wet environment and this absorption plasticises the adhesive thus reducing its modulus.   Again, work with box beams has shown that although discernable, the effects are not so marked as to be of concern in anything other than safety critical structures.   And here, considerable care needs to be exercised in order to determine exactly what is likely to happen.

Mechanical Effects.

It is curious that in view of a general determination to study every conceivable variant of the peel test almost nothing has ever been done to demonstrate how the "Strength", in its widest sense, of an adhesive is enhanced by submitting it to a compression load. Yet it is readily demonstrated that even under impact conditions a toughened epoxide will retain the integrity of a three ply [6,3,3 mm or 3,6,3 mm] laminated steel beam when flexurally loaded during three point impact. The strain in a test like this is considerable and is so great for a two ply [0.125, 0.125 in] laminate that the lower ply is ruptured yet, the rendered metal remains bonded to the upper ply. This is because the adhesive is subject to high shear and compression loads at the moment of impact. This phenonemon was taken into account in the structure of the suspension turret of the ECV3 (the all bonded aluminium quasi monocoque) built by British Leyland Technology (now known as Gaydon Technology Limited) which has been show to perform satisfactorily on both pave and the pot hole breaking tests.

Exhaustive peel and cleavage testing has shown that even the most advanced adhesives, and anything likely to be seen in the near future, are incapable of taking the substantial asymmetrical tensile loads seen when metal deforms and that mechanical reinforcement is required. What is needed now is an undertaking to explore fully the performance and capabilities of adhesives when compressed and shear loaded.

Formulation Changes.

Another problem which must be faced is the change in properties brought about by formulation variations. Generally these will not be important where lightly loaded structures are involved but again, in safety critical areas and power transmission systems, modulus variations and other associated properties may be crucial. Engineers should not ask for a viscosity change, for example, and expect that the change can be made in isolation. It is in areas such as this that the standard test methods can be so misleading. Normal peel and shear tests may show no difference between the old and new materials. But the modulus may have been changed and this should be checked if critical items are involved.

4. DESIGN.

4.1 Load Distribution

It will be obvious from the foregoing that adhesives should never be asked to support asymmetric tensile loads unless these are trivial. In addition, wherever possible potential peel and cleavage forces, induced by accidental or even intended impact, should be anticipated and provision made for

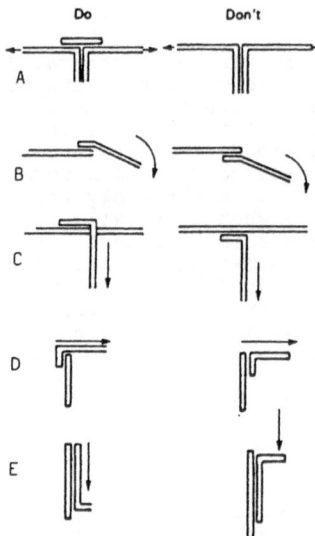

Figure 12.   Acceptable and unacceptable practice in joint design.

their suppression. Mechanical means - welds, rivets etc.
may be used or better still, the shape of the parts
themselves can be employed. A particularly good example of
the latter is the suspension turret of the ECV3, whose design
is based on example 'B' in Fig. 12 where a series of
classical joint forms and force vectors is given. These have
all proved successful and it should be noted that all induce
a compression element in the loading patterns.

The basic load bearing component of any structure is the beam
and often the box beam. Typically, these are made from
separate pieces of folded sheet and one of the simplest forms
is given in Fig 13. Beams such as this and the related
academic design already illustrated in Fig.11 have been
extensively studied. See Table 3. While the all-bonded
equivalent is at least as good, when statically loaded, as
the all welded (Spot) beam, the bonded versions do not behave
as well when impacted on either the end -in line with the
axis, or on the face of the closing plate - at right angles
to the axis. The performance observed in such situations,
when the structure is loaded beyond its capacity, bears no
relationship whatsoever to the performance of a simple bonded
lap joint.

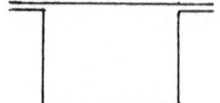

Figure 13.   Beam based on simple
             'Top-Hat' design.

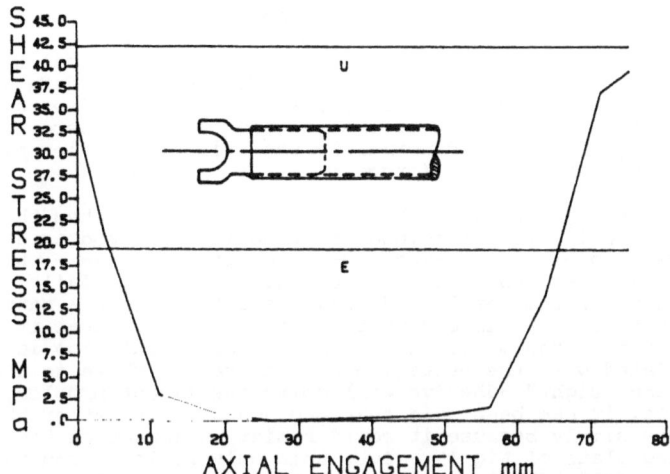

Figure 1⁴. Joint stress distribution generated by a solid drivingyoke bonded into a composite torque tube under a load of 4.5 kNm.

Line U : Ultimate strength of adhesive.

Line E : Elastic limit of adhesive.

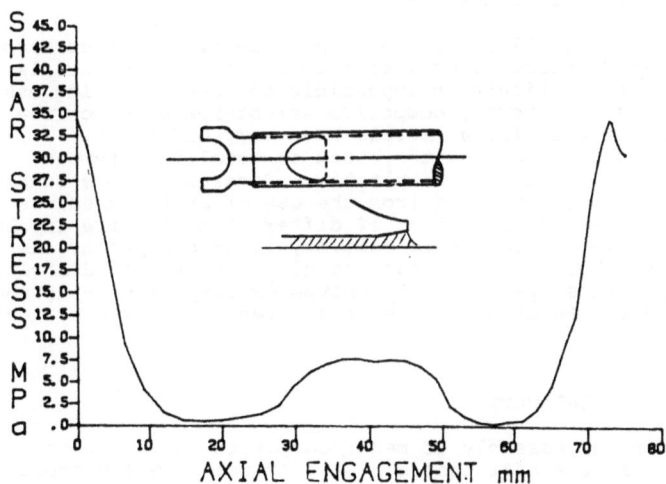

Figure 1⁵. An improved stress distribution given by modification of the yoke's design.

Similarly, the precise data that can be obtained from the 'Thick Adherend Shear Test' cannot be used to predict events because neither compression nor tensile forces normal to the shear axis are taken into account. However, such data are currently used very successfully for the design of bonded power transmission systems (See section 4.3) for here there is no significant load normal to the axis of rotation i.e. no major peel or cleavage forces.

In collapsing box beams the adhesive fails only when the metal itself distorts beyond its plastic limit and in doing so exerts peel and cleavage forces on the adhesive - which, having a lower cohesive strength than the metal, fails. By contrast, if the metal folds and distorts in such a manner that the adhesive is compressed, it is unlikely that it will fail even if the metal itself snaps. Much confusion is associated with the mechanism of failure and there is a view that the "right" adhesive will solve the impact problem. For example, it has been said that one adhesive is better than another simply because it could resist an impact on the closing plate of Fig 13. But, this only follows from the "successful" adhesive being much more ductile than the other adhesive -which only means that at high temperatures it could be far too 'soft' to carry the loads which might be imposed. Whereas, by contrast, at a higher temperature the stiffer adhesive could have just the right ductility and would then resist the impact. On the other hand, at very low temperatures, both adhesives could well be too stiff and both would "fail".

4.2 Dis-similar materials

Adhesives fulfill one of their most important roles when used, as discussed above, to join dis-similar materials which are either difficult or impossible to assemble using other techniques. However, composite structures based on the bonding of dis-similar metals have very different requirements to composites based on totally different materials -say metal and plastic. Generally, more compact designs can be obtained from the use of stiff adhesives in all metal assemblies -even if different metals are involved. But usually, though power delivery is an exception, the more ductile adhesives are needed to give better load distribution when plastics -particularly polyester composites - are bonded to metals. Reference has already been made to this. See Figure 10.

4.3 Power Delivery

It is in the assembly of metal/composite (epoxy, carbon-fibre) drive shafts that adhesive assembly really comes into its own and demonstrates what can be achieved. Relatively free of the problems generated by tensile forces normal to

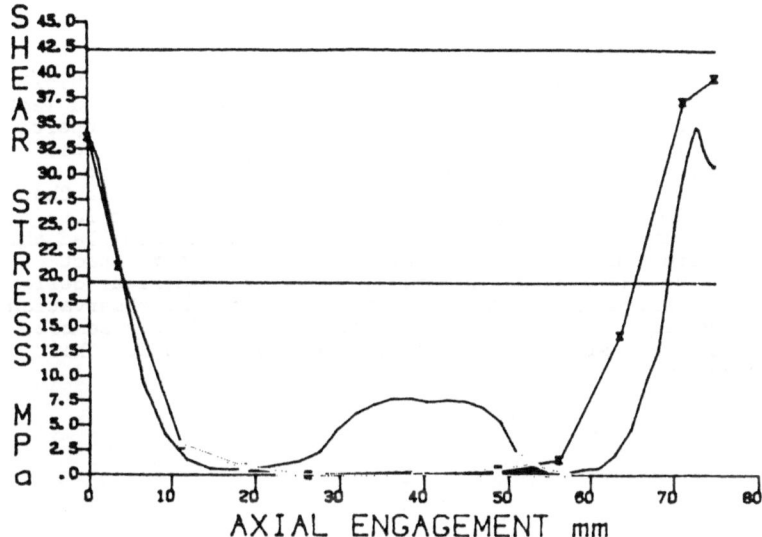

Figure 16. Comparison of stress patterns generated by the two yoke designs of Fig. 18 and 19: Critical edge stress is sharply reduced by transfer of load to central areas - but, creep will still be avoided because of the two unloaded zones. Alternative solutions are bonding both inside and outside the tube, or increasing diameter.

the axis of rotation the correct adhesive functions well and such shafts have outperformed conventional steel components.

Success has been largely related to theoretical design work using mathematical models - of the type developed by AERE Harwell and the ESDU Group - which can cope with elastic/plastic behaviour of the adhesive. As a consequence, it is realistic to believe that such models will be developed further to devise an ideal joint geometry and to identify appropriate adhesive characteristics. A simple example, based on geometric change, is given in Figs. 14 and 15 where the stress pattern induced by a torque load is modified and reduced as the sophistication of a component's form is developed. The modified component is an aluminium yoke bonded into a composite torque tube.

In the first design (Figure 14) the bonded portion of the yoke takes the form of a simple solid bar. In the second (Figure 15), the end of the bar has been hollowed out and the resulting tubular section tapered on both sides. The elimination of stress risers proves to be very beneficial. This may be seen by comparing the stress patterns illustrated in Figure 16.

5. SUMMARY.

Adhesives have much to offer the structural and mechanical
engineer, but their utilisation is being hampered by a lack
of training and knowledge of the true engineering properties
of the materials involved. The former is receiving some
attention though tremendous efforts are being made to
overcome the latter and develop computer based models that
will allow both conceptual and detailed design. So far, only
relatively simple mechanisms can be adequately modelled and
the performance of large bonded structures under impact
cannot be anticipated in detail. However, empirical design
rules derived from both modelling and practical observation
are available and have been successfully used to assemble
load bearing structures which rely on adhesives for their
integrity.

AN ADHESIVE INTERLEAF TO REDUCE STRESS CONCENTRATIONS
BETWEEN PLYS OF STRUCTURAL COMPOSITES

Raymond B. Krieger, Jr.
Mgr, New Developments, Adhesives
American Cyanamid Co., Engineered Materials Dept.
Old Post Road, Havre de Grace, MD 21078

1. INTRODUCTION

The advent of composite laminates on the airframe scene has not always
led to success commensurate with the high promise of material
properties. In many secondary structures, the designs have saved
weight and money. In primary structure, glass and Kevlar have, in some
measure, given way to the superior properties of carbon fibre
composites. The simpler designs in carbon fibre have generally met their
structural goals. In more complex designs, premature failures have been
unexpected and disconcerting. The reasons for these failures are not to
be found in the fundamental properties of the laminate, i.e., tension,
compression, flexure, and inter-laminar shear. The author suggests that
a primary cause is the Achilles heel of the transfer of shear due to
bending between the skins and stringers of "box" beams in wings and
empenages. The same problem waits in fuselage structure using skin and
stringers. These shears have not been wholly overlooked by stress
analysts. Plies of adhesive film are often introduced at strategic
locations where high shear flows are predicted. Nevertheless, the
author suggests there are three questionable facets in the contemporary
approach, namely,
1. While shear flows are readily calculated, the shear stress concentrations are not, because they depend on resin stiffness in carrying
the shear flow transfer.
2. The adhesive ply will mix with the matrix resin. This increases the
adhesive stiffness, and so raises the shear stress.
3. The hot-wet compression strength is reduced because the modulus of
the matrix resin is lowered by mixing with the adhesive.

This paper addresses these problems by A) introducing a new adhesive
film which will not mix with the matrix, and B) offering stress analysis
techniques to more accurately calculate the shear stress concentrations
in order to better predict the strength of the design.

2. CONCLUSIONS
1. It is possible to drastically reduce shear stress concentrations at
critical locations in composite structure. This is done by introducing
a ply, or interleaf, of adhesive film, with low stiffness, at the
strategic location.
2. It is possible to predict the shear stress concentrations because
the requisite adhesive shear stiffness can be known and the stress
formulae have been developed.
3. It is possible to obtain maximum reduction of shear concentration
and minimum loss of hot-wet compression strength by use of an adhesive
formulated so as not to mix with the matrix resin.

3. SHEAR TRANSFER IN LAMINATES

In beams made from laminated plies of composite there are often serious shear forces which must be transferred between plies. Since this plane is not directly crossed by fibres of the composite, the shear loads must be carried by the matrix resin. acting as a thin glue line. This resin plane, or glue line. can become overloaded because of shear stress concentrations. Shear failure on this plane can preclude the structural test or even trigger the collapse of the entire beam. well below design strength.

To understand the situation more clearly, we can begin with Fig. 1. This shows a square increment of laminate loaded with a shear flow around its edges. The shear flow is equal and opposite on parallel sides, and causes a shear deformation from a square to a diamond shape. Moving to Fig. 2, the thickness of the laminate has been abruptly increased over half of the increment. The shear flow on the thinner half must now redistribute over the extra laminate thickness. This results in a sudden concentration of shear force at the edge of the thicker half. This can readily overload the matrix and cause splitting between the plies, as shown.

To visualize the shear concentration problem, it may help to add a layer of adhesive at the critical plane. Fig. 3 shows this glue line as far thicker than the matrix resin layer between plies. It is seen that the adhesive is deformed (loaded) at a maximum at the edge and fades to zero at some point when the laminate shear strain is equal through its entire thickness.

4. SHEAR FLOW in A "BOX" BEAM

Fig. 5 is a schematic drawing of a possible box beam serving as the bending structure of an aircraft wing. The skins and spar webs carrying a shear flow which ultimately produces the tension and compression loads in the stringers and chords. A simplified picture of these shears is shown in Fig. 5. It is noted that the shear flow is variable around the box perimeter. The flow is a maximum in the spar webs and drops each time a chord or stringer is passed. The drop is the amount of shear needed to load the stringer intension or compression. If we look closely at a typical stringer we may see the shear concentrations that can prematurely destroy the structure. Fig. 6 shows the deformations due to shear transfer at a stringer. The shear peaks (concentrations) are shown at several typical locations.

5. Quantitative CALCULATION OF SHEAR STRESS CONCENTRATIONS

In Fig. 3 it can be seen that the shear load transfer in the adhesive is stiffness driven in the same sense as in the classic metal-to-metal bonded skin doubler specimen, Ref. 1. Fig. 4 shows this specimen deformed when the skin (or laminate) is under tension. The adhesive shear stress distribution is shown. The peak stress is given by the equation

$$\frac{KP}{\sqrt{t.t_a.E/G}} \qquad (1)$$

where K = 0.7, P = load per inch, t = laminate thickness, t_a = glue line thickness, E = laminate tensile modulus, and G = adhesive shear modulus. (Ref. 1)

Fig. 3 shows the same principle for adhesive shear stress distribution. Equation (1) will apply since all the stiffness parameters are the same, except of course that E (laminate tensile modulus) must now become GL (laminate shear modulus).

6. REDUCTION OF SHEAR PEAK BY USE OF AN ADHESIVE INTERLEAF

The addition of the glue line, shown in Fig. 3, is a great improvement in reducing shear stress concentrations. It is readily seen by inspection of equation (1) that the glue line stiffness is a function of the glue line thickness (ta) and the shear modulus of the adhesive (G). This makes it possible to closely estimate the improvement percentage, or ratio, obtainable by adding the adhesive interleaf. For Fig. 2, no adhesive, the matrix shear modulus, G, is 200,000 psi and the "glue line thickness" of the single layer of matrix between plies is 0.00004 inches.

Then $\dfrac{1}{\sqrt{t_a/G}}$ = 70,900 = stiffness factor (2)

Next, for Fig. 3, adhesive interleaf, t_a =0.005 inches and G = 100,000

Then $\dfrac{1}{\sqrt{t_a/G}}$ = 4460 = stiffness factor (3)

The shear stress reduction is proportional to these stiffness factors, and so the reduction is 70,900/4460 = 16 to 1.

162

It is recognized that equation (2) may be too stringent. The "stiffness factor" may be too high. More than one layer of matrix resin between plies is deforming in shear. Let us assume a layer of composite several plies thick, acting as adhesive at the critical plane. We may say ta =0.005 inches and G = 500,000 psi

so then $\dfrac{1}{\sqrt{t_a/G}}$ = 10,000 (4)

This factor may at least be doubled since there will be a gradient of shear strain peaking at the center of the 0.005 inch "glue line" so, we may say that the shear stress reduction is 20,000/4460= 4.5 to 1. It is noted that, in equations (2), (3), and (4), the values for adhesive shear modulus, matrix shear modulus, and composite shear modulus were measured by KGR-1 extensometer (Ref. 2).

Equations (2), (3), and (4) are for the linear range of shear modulus. This means that the stress reductions apply to the lower load range of fatigue and creep. For ultimate load, the matrix modulus will not change substantially, but the adhesive exhibits very high ultimate shear strain. The "effective modulus" could be 1/3 of the linear modulus and the stress reductions as high as 27 to 1 and no lower than 8 to 1.

These calculations are seen to have great scatter and to be based on considerable assumptions. Nevertheless, we offer that the minimum improvement of 8 to 1 with the adhesive interleaf is entirely convincing.

REFERENCES

1 "Evaluating Structural Adhesives Under Sustained Load in Hostile Environment", Oct. 1973, 5th National SAMPE Conference.

2 "Stiffness Characteristics of Structural Adhesives for Stress Analysis in Hostile Environment", Oct. 1975, American Cyanamid Co.

Copies of each of these papers are available from American Cyanamid Company, Engineered Materials Dept.,Havre de Grace, Maryland.

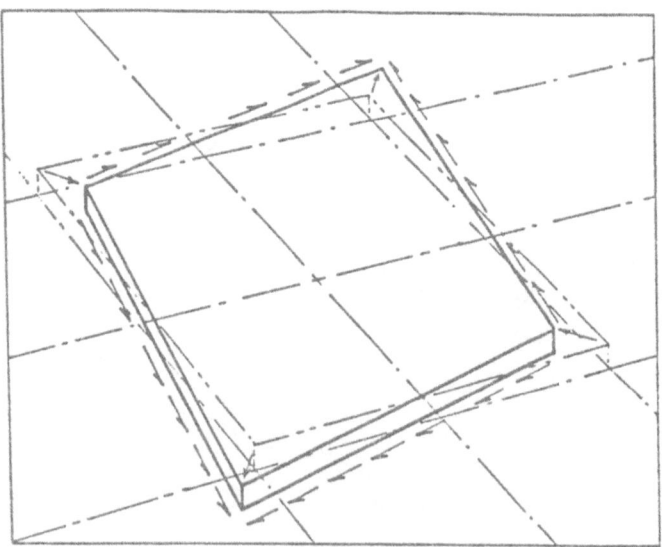

Fig. 1 Shear Flow and Deformation on Increment of
        Composite Laminate

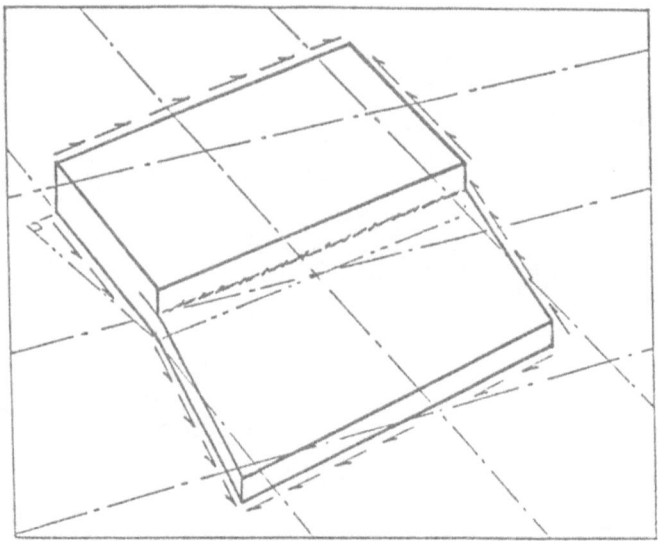

Fig. 2 Shear Flow and Deformation at Thickness Increase
        in Laminate

164

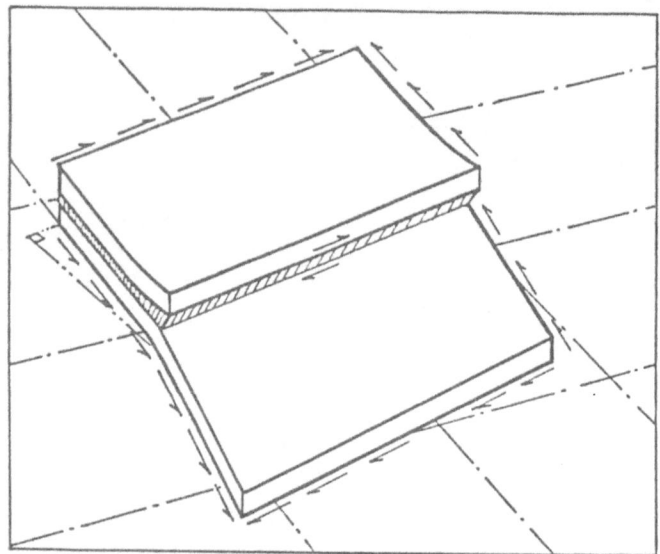

**Fig. 3   Adhesive Shear Distribution When Laminate
Increment is in Shear**

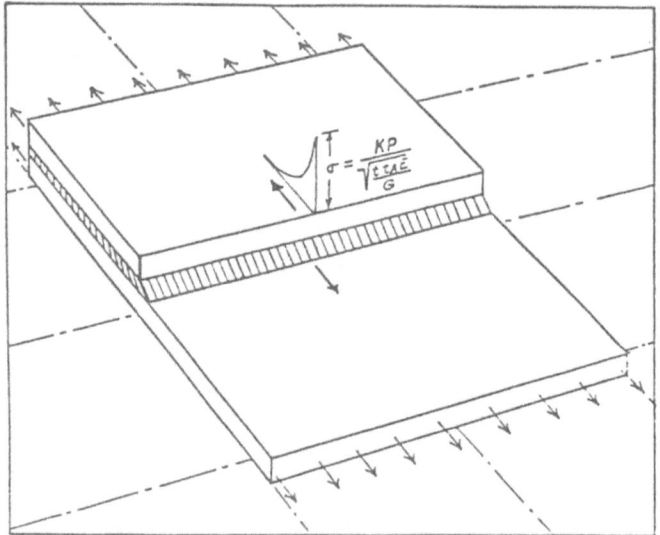

Fig. 4   Adhesive Shear Distribution when Laminate
Increment is in Tension

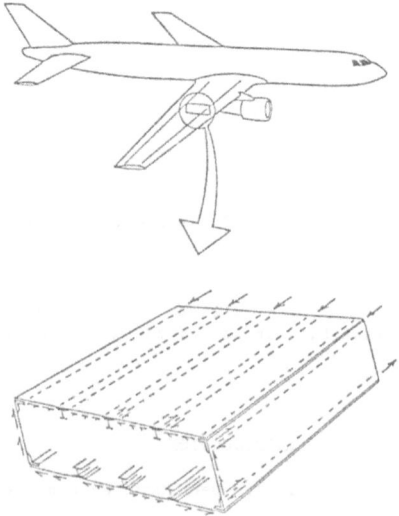

Fig. 5 Wing Box Beam Showing Shear Flow Producing Bending Stresses

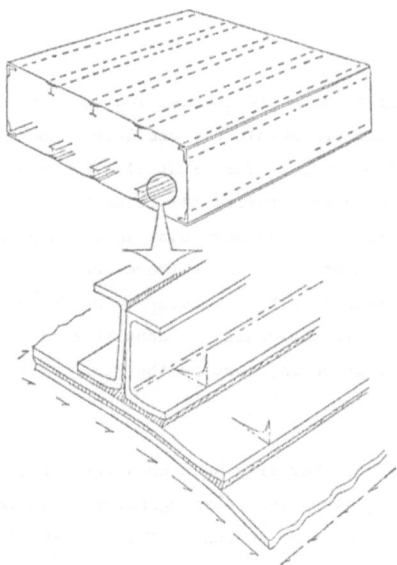

Fig. 6 Adhesive Shear Strains Showing Stress Concentration Points at
a Typical Stringer

CHEMISTRY OF PHENOLIC AND EPOXY ADHESIVES

H. Kollek
Fraunhofer-Institut für angewandte Materialforschung

INTRODUCTION

Phenolic and epoxy resins have been used as adhesives for several decades. They are the most common materials in this field. The curing mechanisms of these resins seem to be very well known and in every textbook of adhesive joining there is a chapter which describes the chemical reactions of curing. New investigations on this chemistry should not give any better knowledge.

But on the other hand, every chemist learns that there are hardly any chemical reactions, which are absolutely complete. Most reactions run to an equilibrium and will then stop. This must also be true for the formation of a polymer, because with the growing molecules their mobility and reactivity is strongly decreased. Another point is the appearence of side reactions in organic chemistry. Mostly there are different ways in which compounds can react with each other, leading to different products. It seems to be obvious that in the curing of adhesives, side reactions will also occur. So the question is, what is the yield of the polymerisation of an adhesive?

These thoughts, together with some unexpected results in the behaviour of adhesive joints, were the starting point of an investigation of the curing of phenolic and epoxy resins. The description of side reactions requires a good knowledge of the base reaction, and the decision was made to investigate both. The work was limited to phenolic resins basing on phenol and formaldehyde and epoxys cured by amines.

THE PRECONDENSATION AND CURING OF PHENOLIC RESINS

Phenolic adhesives are based on a polymer formed from phenol and formaldehyde. Such a resin is produced by a base catalysed reaction leading to a resol. This precondensation leads to a large number of different compounds, which can be analysed by gas chromatography coupled with mass spectrometry, Fig. 1 [1].

The reactive sites of these molecules are the o- and p-position of the aromatic ring, if it is not blocked already by a methylhydroxy group, which is the reactant. The ideal polycondensation needs a ratio of one to one. Theoretically all methylhydroxy groups will then become methylene bridges between two rings giving the highest possible rate of crosslinking. This principle does not take account of the catalyst. In a basic medium the first reaction, the addition of formaldehyde to the phenol is preferred. The polycondensation to polyaromates is quite slow. So most resols consist principally of the compounds in Fig. 1 and there are only traces of higher condensation products. Using an acidic medium the formation of the polyphenols (novolaks) over methylene bridges is faster and hardly any methylhydroxy groups can be found.

As a result both reactions do not lead to a very reactive phenolic resin. A resol has a low concentration of free reactive ring positions and a novolak a low concentration of methylhydroxy groups. By neutralizing the resol after the precondensation, the further polycondensation product will be more like a novolak, but normally the neutralization will not lead to an absolutely neutral product, which depends on the accuracy of this step. From this it becomes obvious that the acidity of a resol will rule the properties of the cured phenolic adhesive. The curing of a phenolic resin will be more or less the formation of a novolak depending on the acidity.

Besides curing an adhesive also has to adhere. Phenolic adhesives are able to form a relatively strong chemical adhesive bond. This cannot be made by the phenol alone, but o-methylhydroxyphenols are able to form complexes with the oxide on a metal surface [2-4]. So a high concentration of resol in a resin will promote adhesion, which only can be obtained if the rate of precondensation is not too high. On the other hand, curing

requires methylhydroxy groups and converts these into the methylene bridges,
lowering the adhesion properties of the cured phenolic resin.

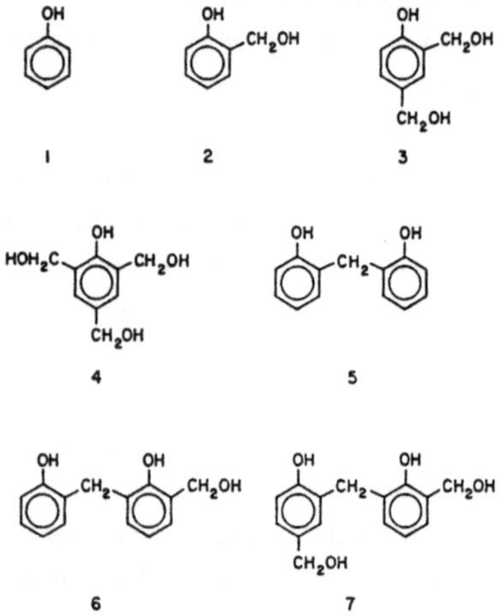

Figure 1.   Typical compounds in resols:

1.   phenol
2.   hydroxymethylphenol
3.   bishydroxymethylphenol
4.   trihydroxymethylphenol
5.   dihydroxydiphenylmethane
6.   hydroxymethyldihydroxydiphenylmethane
7.   bishydroxymethyldihydroxydiphenylmethane

The acidity of a resol can be measured easily by shaking a small
quantity with a few milliliters of water and measuring the pH with test
strips.   The optimum pH range is between 6 and 8.5.   Outside this range
the mechanical properties may be decreased strongly.   Dependence on the
type of catalyst or on the neutralizing agent was not observed.

THE REACTION OF EPOXIDES WITH AMINES

The chemistry of curing of epoxide resins with amines cannot be
analysed very easily, because the crosslinking produces an insoluble
product while most analytical methods require a solution.

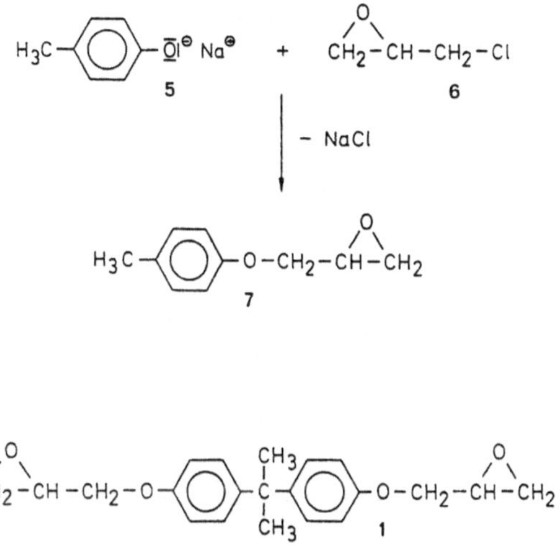

Figure 2.   Preparation of the monoepoxide and
              comparison with the bisphenol A
              epoxide.

In this case a model reaction can be quite helpful, by using a
cresol epoxide made from o-cresol and epichlorhydrin.   This compound can
be understood as "semi bisphenol-A-epoxide", Fig. 2 [5,6].   The reaction
products of this compound with amines are not polymers and therefore can be
analysed easily, e.g. by thin layer chromatography, NMR or mass spectroscopy [5,6].   Here only one example will be described in detail.

It is the reaction of diethylene-triamine (DETA) with this mono-
epoxide.   This amine is very similar to triethylene-tetraamine (TETA)
which is often used as a curing agent for epoxy adhesives.   DETA contains
two different amino groups.   At the ends of the molecule there are two
equivalent primary amino groups and in the centre there is a secondary one.
In a reaction with two moles of the monoepoxide and one mole DETA, the
reaction has finished after 1 day.   The only reaction product is a
symmetric adduct, in which the primary amino groups have been converted to
secondary β-aminoalcohols, [29] Fig. 3.   In principle there are now five
functional groups with three different reactivities, which are able to

react with further epoxide.   The protons of the OH-groups or the NH-protons of the β-aminoalcohols can be substituted by the epoxide or the secondary amine in the centre can be a reaction partner.

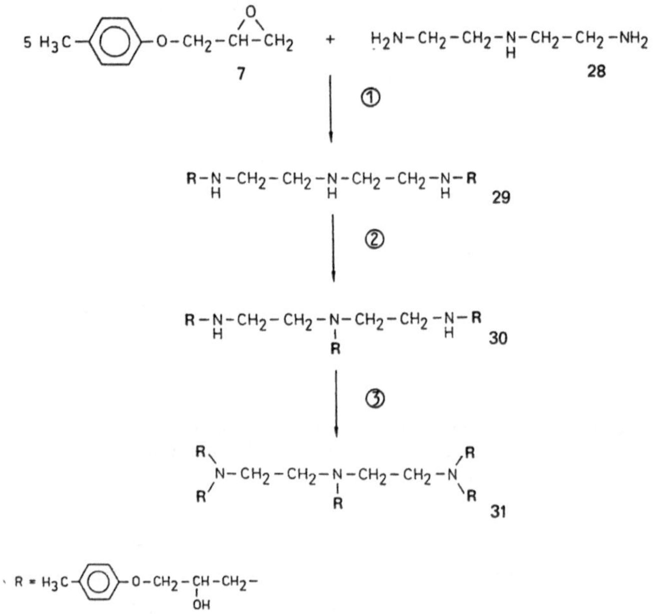

Figure 3.   Stepwise addition of DETA to mono-epoxide.

Increasing the molar ratio of the epoxide to DETA to 3 to 1 and the only reaction product is the product in which all amino groups of the DETA have reacted [30] Fig. 3.   A further reaction of the β-aminoalcohols or asymmetric products are not detectable.   This reaction has finished after 2 days at room temperature.

At a molar ratio of 5 to 1 the penta-adduct [31] Fig. 3 can be detected.   This needs a time of nearly 5 days.   In this compound all protons of the amino groups have reacted with the epoxide.   At room temperature the hydroxy groups do not participate in the reaction with the epoxide [7].

These results from the model reaction have now to be transferred to polymerising systems. The first fast reaction is the formation of a linear polymer. The next two steps lead to a crosslinking and they are much more slower than the first one. On the other hand, polymerising leads to a very low mobility of the molecules and as a consequence to a low reactivity of the system. The formation of the linear polymer in the first step hinders the following crosslinking, which becomes more or less a matter of accident. So it can be proposed, that most of the cold curing epoxide adhesives have only a low rate of crosslinking.

The crosslinking to a certain extent, rules the mechanical properties. From this, the often observed postcuring at elevated temperatures can also be explained. The mobility of the molecules will rise and there are more possibilities for a further crosslinking. Depending on the temperature, the amines then can also act as a catalyst and the hydroxy groups can take part in the reaction and polyethers are formed.

These results from the chemical investigations should also have a strong influence on the mechanical behaviour of the polymer. So the results of two different mechanical experiments were compared with each other |8|. One test was made with the dog-bone specimen and the other one was a single lap shear specimen made from shot blasted mild steel. The resin bisphenol-A-epoxide and DETA were used to form the resin. The molar ratios were 1:1, which leads to a nearly linear polymer and 1.5:1, which allows the crosslinking over the central secondary amino group. The resin was cured at room temperature.

The dog-bone specimens were very brittle and had Young's modulus of 3,125 N/mm^2 and 2,590 N/mm^2 respectively. The tensile strength was 24 N/mm^2 and 44 N/mm^2. The Young's modulus is of the same magnitude as for a plasticised commercial adhesive, but the strength is clearly lower. The strains in the model resins were less than a tenth of that of a commercial adhesive. This shows quite clearly how brittle this material is. It is to be expected that it would be a very poor adhesive, which needs good deformation properties and high strength. The results of the lap shear specimens made from a mild steel are 5.7 N/mm^2 for both concentrations of the DETA, which is quite low. Although the two polymers have quite diffe-rent mechanical properties in the dog-bone specimens, the lap shear strength does not differ. Especially the polymer with the molar ratio of 1:1 was extremely brittle in the bulk material and seemed to have almost no defor-mation properties, so there should be a difference in the lap shear speci-mens. An explanation for this behaviour may be a certain higher order in the bulk material.

SUMMARY

The chemistry of curing of phenolic and epoxide resins strongly influences the mechanical properties of these materials, especially if they are used as adhesives.

For phenolic resins the acidity of the resol is of great importance for the final curing and for the adhesion properties. In the curing a partial convertion of methylhydroxy groups to methylene bridges takes place. On the other side the hydroxyls are essential for the formation of complexes with metal surfaces and for the adhesion properties. Good properties are only achieved if the crosslinking is not too high, which is ruled by the acidity of the resol.

The curing of epoxides with amines runs stepwise. It begins with the formation of a linear polymer which is later crosslinked to a three dimensional network. It can be assumed that the rate of crosslinking is quite low if the resin is cured at room temperature.

ACKNOWLEDGEMENT

The author appreciates gratefully the assistance of Prof. H. Brockmann and Dr. H. Mueller von der Hagen from the University of Bielefeld, Prof. F. Vohnwinkel and R. Bußmann from the Fachhochschule Osnabrück and Dr. A. Groß from the Fraunhofer Institut.

REFERENCES

1. Kollek, H., Brockmann, H. and Mueller von der Haegen, H., Int. J. Adhesion and Adhesives 6, 1986, p.37.

2. Kollek, H., Int. J. Adhesion and Adhesives 5, 1985, p.75.

3.   Kollek, H. and Brockmann, W., in G. Bartelds and R.J. Schliekelmann (Eds.), Progress in Advanced Materials And Processes, 1985, Amsterdam, p.83.

4.   Brown, N.M.D., Meenan, B.J., Affrosman, S., Pethrick, R.A. and Thomson, B., Surf. and Interface Anal., 1987, 10, p. 184.

5.   Groβ, A., Brockmann, H. and Kollek, H., Proc. ASE Conference, London, 1986.

6.   Groβ, A., Brockmann, H. and Kollek, H., Int. J. Adhesion and Adhesives 7, 1987, p.33.

7.   Groβ, A., Modellreaktionen zum Aushärtungsverhalten von Epoxidharz-Klebstoffen, Ph.D. thesis, Bielefeld 1987.

8.   Vohwinkel, F., Buβmann, R. and Kollek, H., unpublished results.

THE EFFECT OF THE ADHESIVE THICKNESS ON THE STRENGTH OF A BONDED JOINT.

A.D.CROCOMBE
Department of Mechanical Engineering
University of Surrey
Guildford, Surrey, GU2 5XH

and

A.C.MOULT
Crescent Consultants
University of Nottingham

INTRODUCTION

Adhesives have been used in structural applications for many years now. With most modern high peel strength adhesives joint failure in metal to metal bonds will invariably occur through a cohesive failure of the adhesive. Thus it is a reasonable premise that analysis based on the material properties of the adhesive and adherend should provide a good tool for joint design.

Potentially such analysis offers a means of establishing stress levels within the adhesive, of assessing the suitability of a proposed configuration and of investigating the mechanisms governing failure of the joint. However, there are a number of restrictions and difficulties which prevent this potential being fully exploited. These difficulties are largely concerned with the complexity inherent in the analysis of a non-linear bi-material system with geometric discontinuities. There are essentially two main categories of analyses used; the closed form analysis and finite element (FE) analysis.

The problem with the closed form approach is that the assumptions that are made in order to model the joint invariably considerably restrict the range of its usefulness. This is nowhere more apparent than in the way that the predicted strength of a joint varies with the adhesive (layer) thickness and this will be considered in the next section.

The FE method does not suffer this restriction but does require an extensive amount of both effort and understanding to implement. There are however software packages available now that considerably reduce the effort and this technique has been the principal tool in this work. Currently one of the major difficulties with the FE approach is to apply the results to give predictions of joint strengths. This has been done through the use of appropriate adhesive failure criteria. However most of this work has been carried out with the joints having a single adhesive thickness. There is good reason to believe, as will be shown later, that if applied to other adhesive thicknesses, these failure criteria would not be so successful.

The purpose of this work was to investigate how the adhesive thickness effects the strength of a lap joint. This was achieved through programmes of test and non-linear analyses of lap joints. The results from the analyses have illustrated how failure occurs in the joint and how this failure is effected by the adhesive thickness This in turn has led to the development of a new failure criteria for bonded joints.

BACKGROUND

In the previous section it was stated that the closed form approach of analysis was not capable of assessing the effect of adhesive thickness on the strength of a joint. Fig 1 based on a closed form analysis of a single lap joint [1] at a load of 400 Nmm^{-1} shows that the stress levels increase with decreasing adhesive thickness and, in the absence of any further information, this clearly implies that the joint strength should decrease with decreasing adhesive thickness In practice this seldom occurs as will be evident from the experimental work reported later. This thickness dependency is not peculiar to the particular analysis cited. Other analyses of this and other joint configurations exhibit similar characteristics.

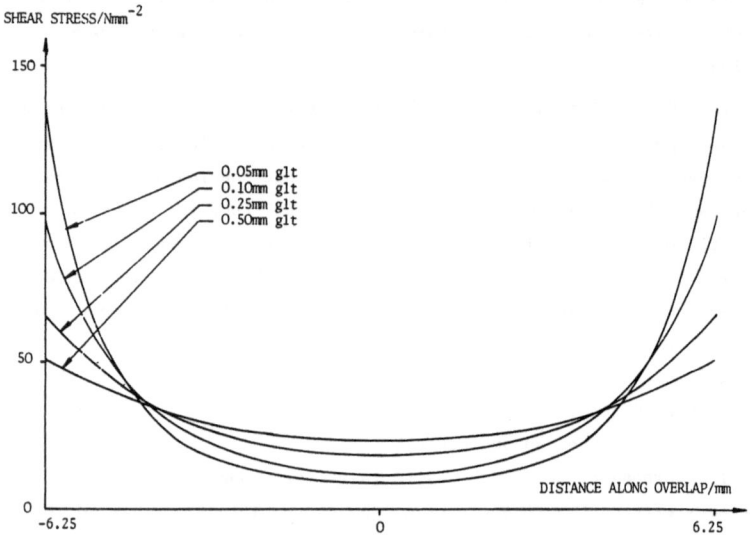

Figure 1. Closed form adhesive shear stress distributions

Although the adhesive shear stress was illustrated in Fig 1 the same
characteristic is noted in the adhesive transverse direct (peel)
stress. In fact much of the work of closed form analyses can be
summarized by the simple formulae presented in Fig 2 giving the peak
peel stress caused by the various load conditions. These formulae were
derived in a separate programme of research [2] concerned with providing
information for joint design. If all other joint parameters are held
constant β_1 and β_2 are inversely proportional to the adhesive thickness
and substituting into the design formulae shows once again that the
adhesive stress increases with decreasing adhesive thickness.

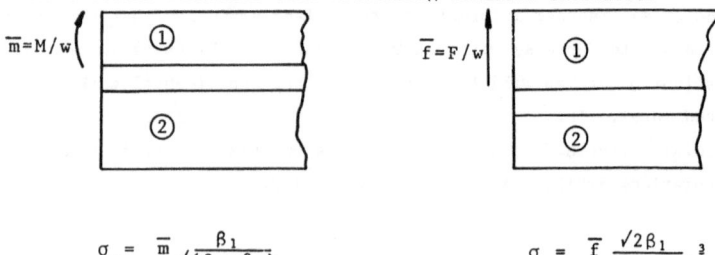

$$\sigma_p = \overline{m} \sqrt{\frac{\beta_1}{(\beta_1+\beta_2)}} \qquad\qquad \sigma_p = \overline{f} \frac{\sqrt{2\beta_1}}{(\beta_1+\beta_2)^{\frac{3}{4}}}$$

Figure 2. Peak stress formulae for joint design.

Fig 3 quickly dispells the possibility that the anomaly between
joint strength and adhesive thickness may in some way be due to the
fillet of adhesive formed at the end of the overlap region. The
adhesive peel stress distributions illustrated in this figure have been
obtained from an elastic FE analysis of the two joint configurations
considered in more detail later in this work, at a load of 400 Nmm^{-1}.
The adhesive fillets simply reduce the level of the peak stress by
spreading the load transfer between the adherends. This feature has
been noted in an earlier publication [3] detailing an in depth study of
the effects of the adhesive fillet on lap joint adhesive stress
distributions. From the figure it can also be seen that accounting for
the joint rotation that occurs by using a large displacement analysis
(first reported in [4]) merely serves to further reduce these stress
levels without changing the thickness effect. The level of stress
reduction correlates very well with data reported in [5].

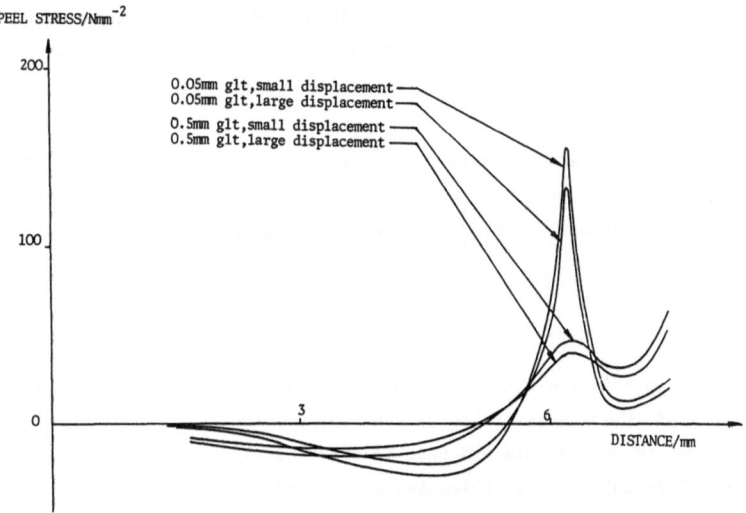

Figure 3. FE adhesive peel stress distributions

Both the closed form and FE analyses reported above assume elastic
material properties and thus it is possible that it is this assumption
that gives rise to the anomaly between joint strength and adhesive

thickness. This is a conclusion that is generally held and it is necessary to carry out a fully non-linear plastic analysis to achieve realistic predictions of joint strength. This approach has already been used in the analysis of the peel test [6].

The double lap joint has been analysed with reasonable success [7] by modelling adhesive plasticity while the adherends remained elastic. Improved joint strength predictions for various adhesive thicknesses were obtained but, even here, the thicker joints were predicted to be stronger. Assuming adherend elasticity is probably a reasonable assumption for double lap joints which are not subject to excessive joint deformation, although recent work [8] shows that considerable plastic deformation in the adherend can still occur.

Both adhesive and adherend plasticity were incorporated in a study of single lap joint performance [5]. By assuming various failure criteria good joint strength predictions for a range of adherend strengths were obtained. However only one adhesive thickness was considered and, as will be seen later, it is quite possible that the same failure criteria would not be so successful at other adhesive thicknesses.

The work reported here can be seen as an extension of the work outlined above, modelling both adhesive and adherend plasticity and examining the effect of adhesive thickness on single lap joint performance.

EXPERIMENTAL WORK

To enable the accuracy of any predicted joint strengths to be assessed it is necessary to have measured joint strengths. It was decided to manufacture and test (and then analyse) sets of joints with two different adhesive thicknesses 0.05 and 0.5 mm. The joints were manufactured in batches of twelve, using 2 mm thick aluminium alloy to specification BSL157. This is a high strength material with a proof stress of around 390MPa. These were subject to a standard chromic acid etch and then jig-bonded to the appropriate adhesive thickness using a room temperature curing aerospace adhesive, HYSOL EA 9309. After initial cure the joint set was post cured at 50° for 8 hours.

The joints were then separated and tested in a standard 30 kN,
screw driven testing machine. Load-deflection plots taken from these
tests will be shown later to support a theory of joint failure. From
these plots the failure load of each joint was obtained and these are
shown in figure 4.

FAILURE LOAD/kN

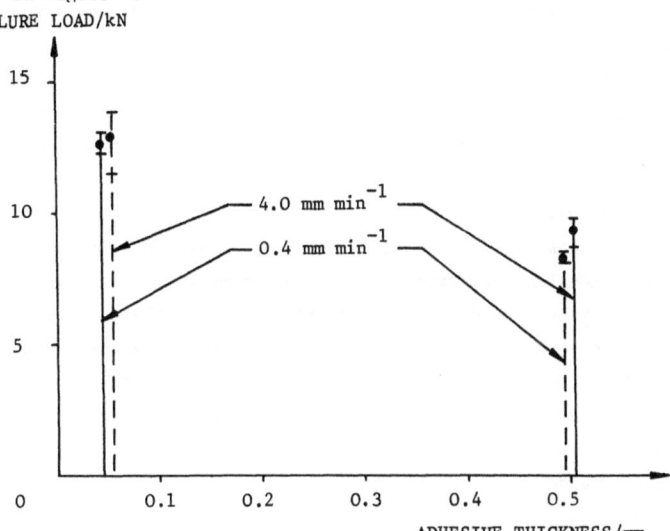

Figure 4. Measured lap joint strengths

In a crude attempt to elimate the strain rate effects that are
likely to arise when testing joints with different adhesive thicknesses,
two different crosshead speeds were used, 0.4 and 4.0 mm/min. By
comparing failure loads from the thicker joint at the faster speed with
those from the thinner joint at the slower speed it was hoped that
similar strain rates would prevail. In any event by comparing the solid
and dashed lines in fig 4 it can be seen that the rate effect is far
less significant than the thickness effect.

The mean failure load for the thinner joint is about 12.5 kN (480
Nmm^{-1}) while that of the thicker joint is about 9.5 kN (380 Nmm^{-1})
Thus precisely the opposite effect to that predicted by elastic analysis
is found i.e. the strength decreases, not increases, with increasing
adhesive thickness The experimental scatter, denoted by the horizontal
bars, can be seen to be small. This is attributed to the fact that in

every joint tested failure occured cohesively within the adhesive,
leaving a clearly visible layer of adhesive on each adherend surface.

THEORY

The theoretical background of the FE method is covered in many
books. Essentially the technique can be considered as a process of
discretisation, subdividing a complex shape into a number of small
regular regions known as "finite elements". By minimising the total
energy of the system an approximate solution for the displacements at
various points in the structure can be found and from these the strains
and stresses evaluated.

Clearly this approach is ideal for the analysis of a bonded joint
having both complex geometric and material properties. It has been used
in this context with success for some time now. As the computer power
has increased so has the mesh refinement used and hence the accuracy of
the solution. Full details of how to extend this approach to include
non-linear behaviour and its application to bonded joint analysis are
available [9] and will not be repeated here except for a discussion on
yield criteria as this will be shown to play an important part in joint
performance.

In defining the yield criteria reference is made to two important
functions of the general stress state:

$$J_1 = (\sigma_x + \sigma_y + \sigma_z)$$

$$J_2 = \tfrac{1}{2}[(\sigma_x - J_1/3)^2 + (\sigma_y - J_1/3)^2 + (\sigma_z - J_1/3)^2] + \tau_{xy}^2 + \tau_{xz}^2 + \tau_{yz}^2$$

J_1 is known as the first stress invariant and is a measure of the
general (hydrostatic) level of stress. J_2, the second stress invariant,
on the other hand, is a measure of the difference in (deviatoric) stress
level. The two yield criteria used are shown in figs 5a and b. Fig 5a
illustrates the von Mises yield criteria which is independent of the
first stress invariant, and thus in principal stress space the unyielded
state is contained within a cylinder whose axis is the line of pure
hydrostatic tension, $\sigma_1 = \sigma_2 = \sigma_3$. Strain hardening can be

accommodated by increasing the radius of this cylinder. Fig 5b illustrates a modified von Mises criteria where yielding is dependent on both the first and second stress invariants, a higher level of hydrostatic stress inducing yield at reduced levels of deviatoric stress. The relative effect of the hydrostatic stress is determined by the ratios of yielding in uniaxial compression to that in tension (S). This has been taken as 1.3 in this work and a justification for this and further information on both yield criteria can be found in another publication [9]

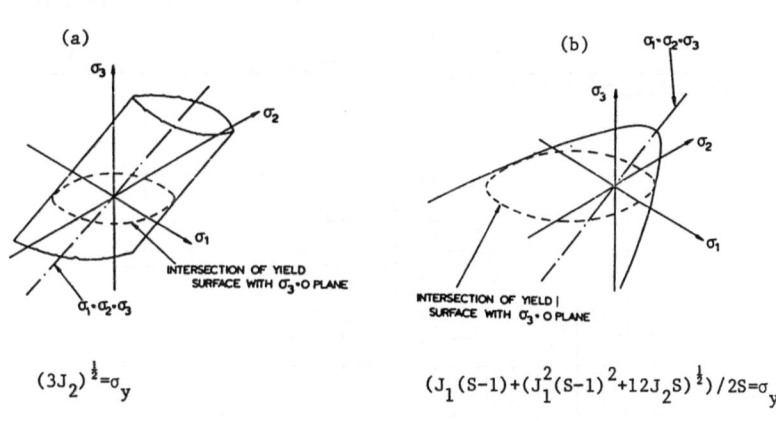

(a)

$$(3J_2)^{\frac{1}{2}} = \sigma_y$$

(b)

$$(J_1(S-1) + (J_1^2(S-1)^2 + 12J_2S)^{\frac{1}{2}})/2S = \sigma_y$$

Figure 5. Various yield criteria

ANALYSIS AND DISCUSSION

The two configurations tested experimentally were subsequently analysed using a finite element package generated during a previous programme of research [10].

The FE meshes used for the analysis of the thicker joint is shown in Fig 6. The mesh used for the thinner joint is the same except of course, that the adhesive elements were thinner. These meshes incorporated 608 elements and 2013 nodes. The elements used were 8 noded quadratic isoparametric elements unless mesh refinement required the use of compatible 6 noded triangular isoparametric elements. The most important features to note are that two elements were used across the adhesive thickness, that successive mesh refinement was used to

concentrate elements along the overlap (particularly at the ends) and
that the adherend mesh was graded towards the adhesive layer. Such
complex meshes can be constructed fairly easily using FE pre processing
packages. In this instance PATRAN, an interactive graphics package from
PDA Engineering was used.

The adherends are constrained and loaded at their ends, as shown in
fig 6, simulating the gripping and loading of the testing machine.

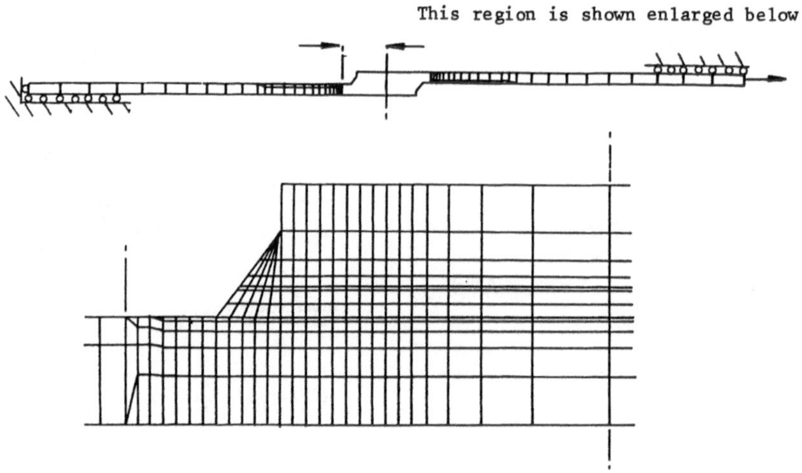

This region is shown enlarged below

Figure 6. Mesh used in the FE analysis

Various material models were used to define both the adhesive and
adherend full non-linear behaviour. These models define the uniaxial
stress - strain curve of the material. The models used for the adhesive
are shown in fig 7. The dashed lines represent the band of experimental
data available while the solid and chain dashed lines show the two
models, a simple bi-linear model and one that accommodates smooth
transition between elastic and plastic behaviour using a second order
polynomial connecting two linear regions. The adherend behaviour was
modelled using a bi-linear model with initial slope of 70GPa and final
slope of 711MPa, transition occuring at a stress of 395MPa.

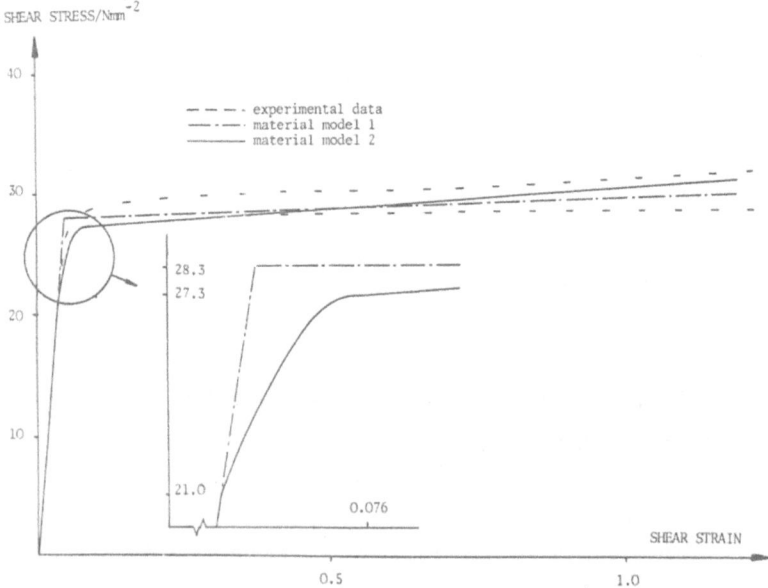

Figure 7. Adhesive material models

Before carrying out a comprehensive set of non-linear analyses it was necessary to validate the results of the FE analysis. This has been done by comparing the results of an elastic analysis of the thicker joint with the closed form analysis of Goland and Reissner [1]. The FE mesh used was the same as that shown in Fig 6 without the fillet of adhesive at each end of the overlap. The initial stage of the closed form analysis is concerned with finding the bending effect of the applied load. This is accounted for by defining a bending moment factor obtained by assuming the adherend to be infinitely long. The appropriate factor to use for the finite adherend length model at the low loads (0.1 Nmm^{-1}) at which this comparison is to be made is 1.14. Using this value the closed form and FE results for the adhesive transverse direct stress are shown compared in fig 8 below. Clearly the correlation is good indicating that the FE model is yielding reliable results.

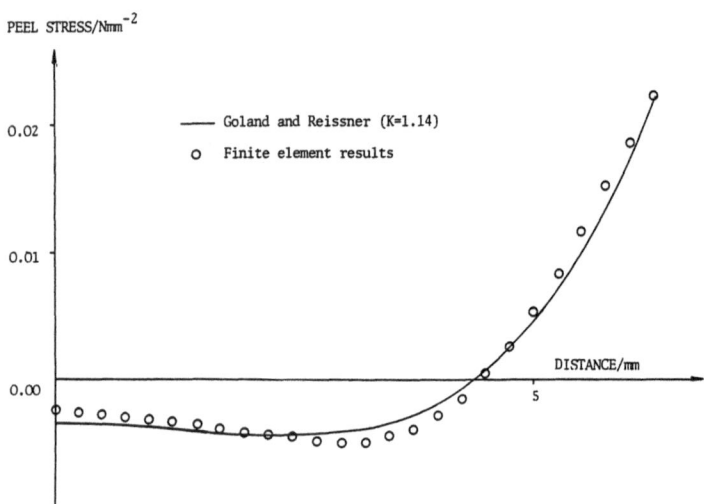

Figure 8. Comparison between closed form and FE results

Two different sets of non-linear analyses have been reported in
this work. The first set were carried out to enable comparison of
results with a second commercial FE package. Due to limitations in the
commercial package yielding in both the adhesive and adherend was
assumed to be governed by the von Mises yield criteria, fig 6a.
Further, bi-linear material behaviour for both had to be assumed and the
effect of joint rotation was not assessed. The second set of non-linear
analyses used the same conditions for the adherend but used the modified
von Mises criteria and the smoother material model for the adhesive, and
accounted for the effect of joint rotation by adopting a large
displacement analysis. The contrast between the two will be seen to be
significant and attributable either to the different adhesive yield
criteria or material model used.

The results of the first set of analyses have been summarised by
comparing the spread of the adhesive yield zone in the thick and thin
joint at various loads. This is shown in fig 9 a - c. Fig 9a refers to
a load of 187 Nmm^{-1} which is sufficient just to introduce yielding into
the thicker joint. Clearly, as elastic analyses suggests, yielding
occurred at a lower load in the thinner joint and is spreading into the
overlap. Initial yield in this configuration occurs under the overlap
region rather than adjacent to the adhered corner in the adhesive fillet

as is the case with the thicker joint. This is a feature that was noted in earlier work [3] and attributable to the relative stiffnesses of the overlap and fillet regions. An interesting feature of the adhesive yielding which has not been previously reported is the spread of the yield zone along the unloaded adherend face in a "thin finger" fashion. This feature is not noted in the second set of analyses and it is felt that this is either due to the material model or yield criteria being used although more work needs to be carried out to determine which.

Figure 9b compares the spread of yielded adhesive at a load of 293 Nmm^{-1}, which lies between initial yield and final failure. From this figure it can be seen that the extent of yielding in the thicker joint is rapidly approaching that in the thinner joint, and that the yield spreads more quickly into the adhesive fillet. There are two possible reasons for the more rapid spread of yielding in the thicker joint:-

a) The elastic analyses considered earlier in the paper show that the level of stress in a thicker joint although lower is spread more uniformly than in a thinner joint. Thus when yielding does occur there is less "elastic reserve" to sustain further loading and thus yielding spreads more quickly.

b) When yielded the adhesive shear stress will be greatest when the direct stresses are in pure hydrostatic tension or compression. This is more likely to be the case in the highly constrained thinner joint and thus for such joints an increase in load can be sustained with a smaller degree of yielding because the shear stress is higher.

Fig 9c shows the extent of yielding at a load of 400 Nmm^{-1}, just above the measured failure load of the thicker joint. Here it can be clearly seen that yielding in the thicker joint is more extensive than in the thinner joint. Further there is nearly a path of yielded material along the whole overlap. When this does occur the lack of strain hardening (flat shape) of the adhesive stress-strain curve means that the joint will not be able to support any further substantial increase of load and gross yielding will occur prior to joint failure.

This theory can be supported by considering the shape of the experimental load-deflection plots, fig 10.

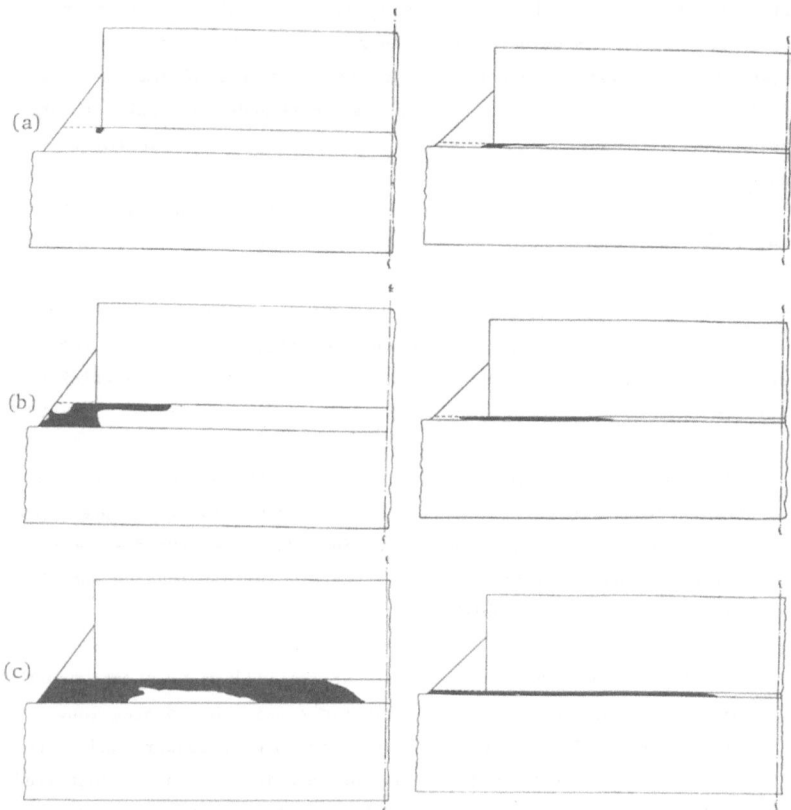

Figure 9. Adhesive yielding at increasing load levels in the first set of analyses

The thicker joint exhibits a region of increase in extension without a corresponding increase in load, this, it is suggested, is the gross yielding, referred to above that occurs when there is a completely yielded path through the overlap. The same feature can be noted but to a much lesser extent in the thinner joint.

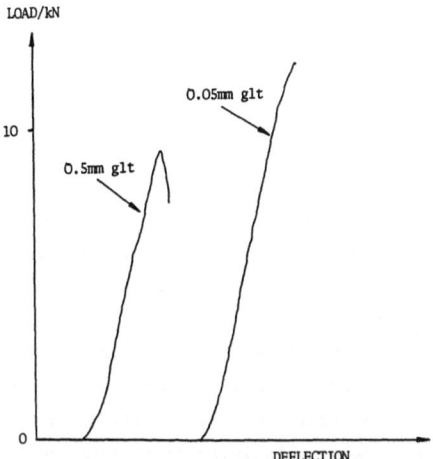

Figure 10. Experimental load-deflection plots

From the FE analyses it is possible to predict the load deflection behaviour and this is shown in fig 11. Once again the region of gross yielding is apparent as a flattening of the load-deflection curve. The thinner joint which has not fully yielded over this load range is much more linear in nature.

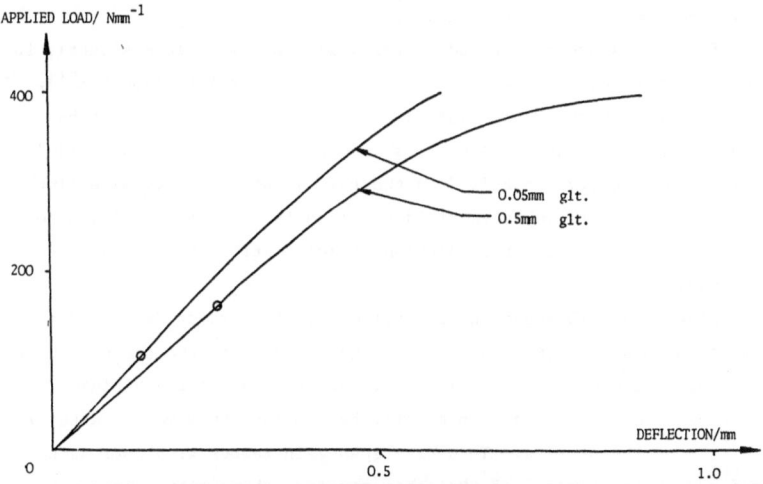

Figure 11. Theoretical load-deflection plots

Thus it can be seen that assuming joint failure to occur when there is a yielded path in the overlap region gives a reasonable measure of joint strength and predicts, <u>for the first time</u> the thinner joint to be stronger.

Results from the second set of analyses are shown in fig 12 a-c. Once again the first of these is taken at a load (164 Nmm^{-1}) just sufficient to induce yield in the thicker joint. Note that even though the stress levels are reduced by accounting for the joint rotation the use of the alternative adhesive material model with lower yield stress (fig 7) results in yielding occurring at lower levels of load. Notice also that the "thin-finger" pattern of yield noted in the previous analyses are no longer present. This is either because the yielding can now also be induced by an increase in the general stress level or because the smoother stress-strain curve initially allows a considerable increase in yield stress over a relatively small range of strain and thus material adjacent to yielded material and at a slightly lower level of stress will soon yield as the yielded material strain hardens rapidly.

Fig 12 b again shows the state of yielding at a load between initial yield and final failure (275 Nmm^{-1}). Once again it can be noted that yielding spreads more quickly in the thicker joint, and that the fillet is largely unyielded in the thinner joint. Also, the thin finger of yielded material is still absent.

Fig 12 c shows the spread of the yield zone at a load slightly in excess of the measured failure load of the thicker joint(400 Nmm^{-1}). It can be seen that while the overlap region for the thicker joint has completely yielded there is still a substantial portion of unyielded adhesive or "reserve capacity" in the thin joint. Thus by modelling the joint rotation, and adhesive yield and material characteristics more accurately an even better prediction of joint strength has been achieved.

Finally Fig 13 shows the effective plastic strains developed at a load of around 260 Nmm^{-1} in both configurations from the second set of analyses. Although, due to yielding, the stresses in the yielded adhesive developed in each joint will be similar the same is certainly not true of the strains. This discrepancy increases with increasing load and, being typical of the other strains, shows quite clearly the

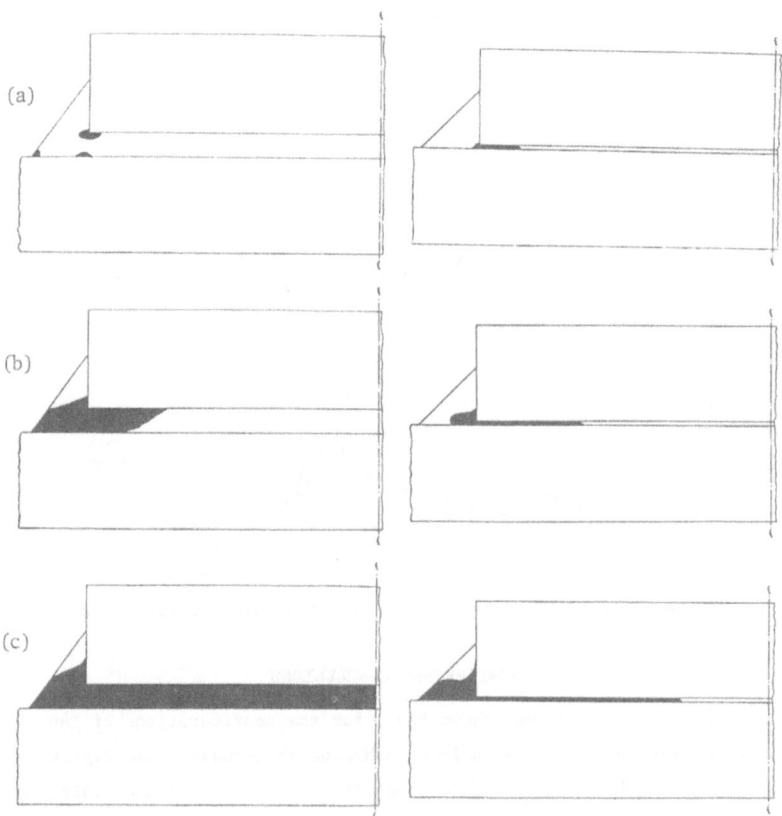

Figure 12. Adhesive yielding at various load levels in the second set
of analyses

difficulty of using a strain based failure criteria to assess the
strength of those joints. A strain based failure criteria clearly
predicting the thicker joint to be stronger than the thinner joint.

EFFECTIVE PLASTIC STRAIN

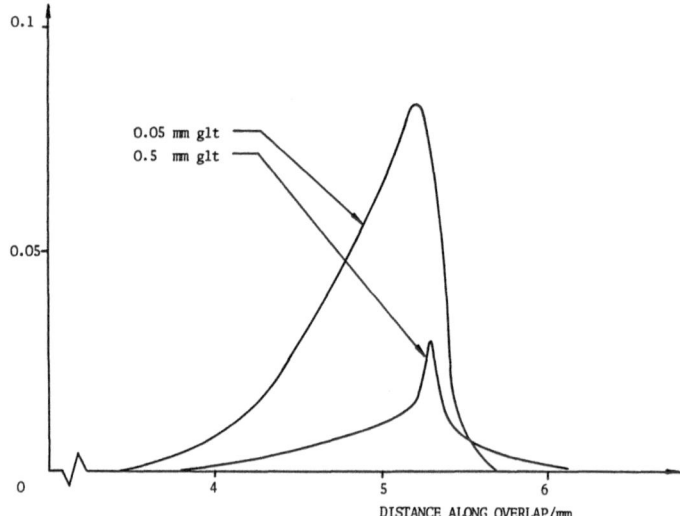

Figure 13. Adhesive effective plastic strain distributions

SUMMARY AND CONCLUSIONS

Experimental work has shown that, for the configurations of lap
joint considered, joints with lower adhesive thicknesses have higher
failure strengths. This is a feature that cannot be explained using any
form of elastic analysis and has not, until now, been explained using
non-linear analysis. By carrying out extensive non-linear analyses into
joints with two different adhesive thicknesses up to the measured
failure load important yielding and failure mechanisms have been
highlighted. Particularly it has been shown that for the configurations
considered a failure criteria based on complete yielding of the lap
joint has given good predictions of joint strength and, for the first
time, has shown thinner joints to be stronger. The use of a strain-
based failure criteria for these configurations has been shown to be
extremely limited. A small change in yield criteria and material
behaviour has produced a significant change in the way the yield zone
spreads, and are aspects which require further investigation.

It seems likely that failure in a bonded joint can occur when at a

point some local criteria for adhesive failure is violated. For the
configurations considered it appears that global yielding occurred
before the local failure criteria was violated. Using this global
yielding as a failure criteria gives a good measure of joint strength.
There will however, be other joints where local criteria are violated
before global yield occurs and this is an aspect that clearly requires
further study.

ACKNOWLEDGEMENTS

The author would like to thank RAE Farnborough and particularly Mr
M. Stone for his interest and support in this work, supplying materials
and background information. Further thanks also go to Mr D. Bigwood for
assistance with the closed form analysis results and graphical
presentation.

REFERENCES

1. Goland M and Reissner E. The stresses in cemented joints.
   J.Appl.Mech., Trans ASME,pA17,1944.

2. Crocombe A D, A general bonded joint peel analysis. Proc.Int. Conf
   on Structural Adhesives in Eng. MEP Ltd 1986.

3. Crocombe A D and Adams R D, Influence of the spew fillet and other
   parameters on the stress distribution in the single lap joint.
   J. Adhesion 13 pp 141-155, 1981.

4. Crocombe A D and Adams R D, Peel analysis using the finite elements
   method. J. Adhesion 12, pp 127-139,1981.

5. Harris J. A and Adams R D, Strength prediction of bonded single lap
   joints by non-linear finite element methods. Int J. of Adhesion and
   Adhesives,4 pp 65-78 1984.

6. Crocombe A D and Adams R D, An Elasto-plastic investigation of the
   peel test. J. Adhesion 13 pp 241-267,1981

7. Adams R D, Coppendale J and Peppiatt N A, Failure analysis of
   aluminium-aluminium bonded joints. Adhesion 2,editor K.W.Allen,
   Applied Science Publishers Ltd 1978.

8. Evans I E J, The effect of adherend cladding and adhesive thickness
   on the failure load of stable lap joints. BEng thesis, University
   of Surrey 1986.

9. Crocombe A D, The non-linear stress and failure analysis of adhesive

tests. PhD thesis, University of Bristol 1981.

10. Crocombe A D, FELDEP - a finite element program for elasto-plastic analysis of structures subject to large displacements, University of Bristol 1981.

BONDING OF COMPOSITES TO METALS

R D Adams
Department of Mechanical Engineering
University of Bristol
Bristol BS8 1TR
U.K.

and

J A Harris
formerly of the above address but now at
Materials Engineering Research Laboratory
Tamworth Road
Hertford SG13 7DG
U.K.

INTRODUCTION

As in any engineering structure, the strength of adhesively-bonded joints depends on the strength of the weakest component. This critical lowest strength may occur in many different parts of the joint and depends on the strength of the adherend, the adhesive, or any intermediate zone between them. In a properly made joint, failure usually occurs in the adhesive (called cohesive failure) or, though rather less often, in the adherends. Rarely does failure occur in the intermediate layers which are at or very near to the interface (called adhesive failure).

The low transverse tensile strength of fibre reinforced plastics (FRP) is well known. Because of this, attention must be paid to the design of bonded joints using FRP adherends so that premature failure in the adherend caused by transverse tensile stress is avoided. Recently, Cushman, McClesky and Ward (1) stated that 'although numerous researchers have investigated the state of stress within a bonded composite joint, few have made an attempt to predict actual failure loads. Also most of these prediction techniques assume a failure of the adhesive, and do not address the problem of interlaminar composite adherend failures'. In this work, finite element methods (FEM) have been used such that joint failure may be based on the conditions in either the adherends or the adhesive layer. FEM enables more complex geometries to be analysed than is possible with closed form methods, such as those by Volkersen (2) and Goland and Reissner (3).

The basic joint under consideration was a double lap joint with a unidirectional carbon fibre reinforced plastic central adherend with steel outer adherends, the dimensions being as indicated in design 1 of Fig 1.

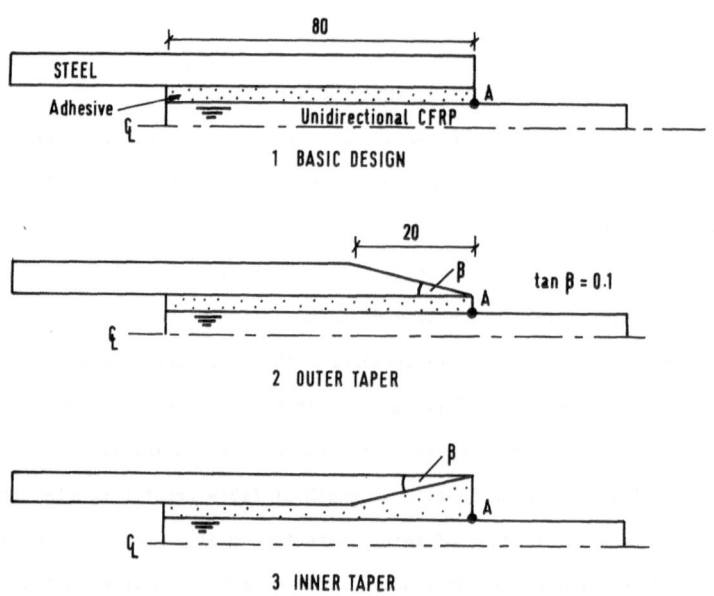

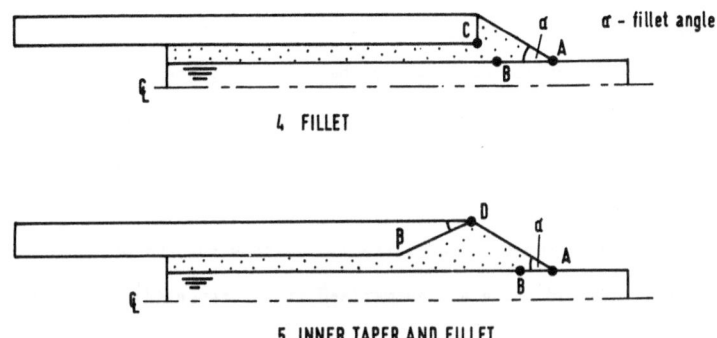

Figure 1. Double lap joint designs

The remaining designs shown in Fig 1 are modifications of the basic design keeping the same overlap but aimed at improving joint strength. In designs 2 and 3, the outer adherends have been modified by tapering; this has been shown by Thamm (4) to reduce the maximum adhesive shear stress in a joint, but only if the taper is continued to a fine edge. Design 4 shows the original joint modified to include an adhesive fillet at the end likely to fail; this has been shown by Crocombe and Adams (5) to reduce the peak maximum principal adhesive stress. Finally, in design 5 both a taper and a fillet have been incorporated.

A toughened epoxy adhesive (Ciba-Giegy XD911) was used. This material has a Young's modulus of 3.05 GPa, a failure stress of 84 MPa, and a failure strain of 4.6 per cent when tested in bulk uniaxial tension.

ANALYSIS

In a lap joint, differential shear in the adherends leads to a shear stress concentration in the adhesive at the edges of the overlap. This situation was first analysed by Volkersen (2). In addition, internal bending moments are set up in the joint and Volkersen accounted for this in his later work (6) so that a distribution of transverse normal stresses exists. The transverse stress shows a maximum value in tension in the adhesive layer at the edges where the outer adherends terminate. The transverse (peel) stresses in this region are very important in assessing joint strength since both the adhesive and the CFRP are weak under this mode of loading.

There are a number of analytical solutions for the state of stress in adhesive joints and these are summarised in reference (7). Although they give a qualitative assessment of the effects of various parameters, analytical solutions do not enable joint strengths to be quantified. A principal reason is that a complete analysis of the various components of stress is required, including variations through the thickness of both the adherends and the adhesive. Also the non-linear properties of the adhesive must be included if realistic materials are to be modelled. Finally, joint strength is significantly influenced by the local geometry in the critical regions of the joint, so it is necessary to account for the existence of a fillet of adhesive at the edges of the overlap, the sharpness of the adherend corner, and so on. Closed-form algebraic (analytical) solutions cannot allow for these factors on the scale necessary for accurate joint strength prediction.

Thus, it is necessary to use the finite element method (FEM) for predicting joint strength. FEM is a well-established mathematical technique for analysing complex situations. Essentially, the continuum is broken down into discrete elements, each of which can be assessed and each of which interacts with its neighbours in a rational fashion. Regions of high stress gradient can be accommodated by using a finer mesh of elements. Material non-linearity and anisotropy can be included in the description of any element. Using the finite element method, it has been shown by Adams and Peppiatt (8), that lap joint failure is initiated by tensile fracture of the adhesive within the fillet close to the adherend corner. They showed that a crack runs through the fillet at approximately 45° to the adherend surface and perpendicular to the predicted directions of the maximum principal tensile stresses in the adhesive. Adams, Coppendale and Peppiatt obtained reasonable predictions of the strength of double lap joints with aluminium adherends (9). Late work by Adams and Harris in which large displacement deformation was modelled gave reasonable predictions of the strength of single lap joints with a range of epoxy adhesives (10) and when using aluminium alloy adherends which could yield extensively.

The adhesive fillet at the edge of the adhesive layer has been shown, using finite element techniques, to reduce the maximum stresses in the adhesive (8). This, together with Thamm's prediction (4) by closed form analysis that, by tapering the adherends to an almost 'razor edge', the peak stresses in the adhesive may be reduced, was utilised in various attempts to improve joint strength as shown for the series of CFRP-steel joint designs illustrated in Fig 1. The modifications to the basic design fall into three categories: firstly, in designs 2 and 3 the outer steel adherends are tapered with a 10:1 gradient and the edge of the adhesive layer is square; secondly, in design 4 the adherends are unmodified, but adhesive fillets, whose size is defined by the angle ' α ', are included; thirdly and finally, in design 5, both the tapered adherends and adhesive fillets are included together.

Almost invariably,the most highly stressed region in an adhesive joint occurs at or near one corner. If the corner is sharp then in theory there exists a singularity which implies infinite stresses and/or strains at this corner. In designs 1-3 the points designated as 'A' in Fig 1 are singularities, as are points 'C' and 'D' in designs 4 and 5. A fracture mechanics analysis may be applied to the stresses in these critical regions. However, the authors know of no successful fracture mechanics method in which the strength of bonded lap joints has been predicted. An alternative approach can be used (11) in which a small degree of local rounding is introduced into the finite element model in the critical region, thus removing the singularity. In this way, the problems of dealing with singularities are avoided and failure criteria applied to the maximum conditions occurring within the predicted stress field may be employed. In practice, the corner geometries are unlikely to be perfectly square anyway, so that the modified geometries are much more likely to be realistic.

Both the steel and CFRP adherends were modelled as linearly elastic materials while, for the adhesive, yield and plastic deformation were taken into account. The yield criterion for the adhesive is a function of the hydrostatic as well as the deviatoric stress component and is of the form (12):

$$[J_1(S-1) + (J_1^2(S-1)^2 + 12\ J_2 S)^{\frac{1}{2}}]/2S = Y_T$$

where J_1 and J_2 are the first and second stress invariants respectively and

$$S = Y_C/Y_T$$

where Y_C and Y_T are the yield stresses in uniaxial compression and tension respectively,

$$J_1 = (\sigma_1 + \sigma_2 + \sigma_3)/3$$

and

$$J_2 = (\sigma_1^2 + \sigma_2^2 + \sigma_3^2)/3$$

and

σ_1, σ_2 and σ_3 are the principal stresses.

The material constants used for the adherends are given in Table 1. The mechanical properties of the steel and CFRP were taken from the literature, and those of the adhesive were determined experimentally by the authors.

TABLE 1
Material Properties

(a) Steel: Young's Modulus   210 GPa
                Poisson's ratio   0.29

(b) Adhesive: Young's Modulus   3.05 GPa
                Poisson's ratio   1.35
                S   1.24

(c) Unidirectional CFRP:

Longitudinal modulus	140 GPa
Transverse modulus	7 GPa
Interlaminar shear modulus	4.5 GPa
Longitudinal and transverse Poisson's ratio	0.3

RESULTS AND DISCUSSION

Stresses in the Composite Adherends

For all of the joint designs considered, the peak transverse stresses, $\hat{\sigma}_T$, in the composite occurred in the region adjacent to the edges of the outer steel adherends. The values of $\hat{\sigma}_T$ for each joint under a load of 1 MN m^{-1} width are given in Table 2. In designs 1-3, tapering the outer (steel) adherends has negligible effect on reducing $\hat{\sigma}_T$. A contour plot of the transverse stresses was produced for the critical region of design 1. Here, as with designs 2 and 3, a large stress concentration exists adjacent to the very edge of the adhesive layer. The abrupt edge of the adhesive layer causes the transfer of the load from the inner CFRP adherend to the outer steel adherends to be focused in this local edge region; the transverse stresses in the CFRP decay rapidly away from this location towards the centre-line of the joint and longitudinally away from the overlap. This pattern of load transfer and stress concentration is affected very little by either the outside or inside taper of designs 2 and 3.

TABLE 2

Predictions of the Maximum Transverse Stresses in the CFRP from Elastic Finite Element Analyses with a 1 MN m^{-1} Applied Load

Design	Fillet angle	$\hat{\sigma}_T$ in CFRP $\frac{}{MPa}$
1	–	38
2	–	37
3	–	36.5
4	45°	16
4	30°	10
4	17°	10
5	45°	13
5	30°	6.5
5	17°	5

By introducing an adhesive fillet in design 4, an appreciable reduction is obtained in $\hat{\sigma}_T$. Even the relatively small modification of a 45° fillet halves the stress. The fillet reduces the focus for the transfer of load at the edge of the overlap, giving a more even distribution of the transverse stress. Figure 2 shows the stress distribution in the CFRP for a full depth 30° fillet.

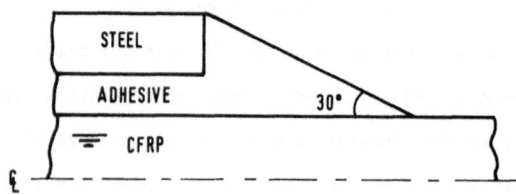

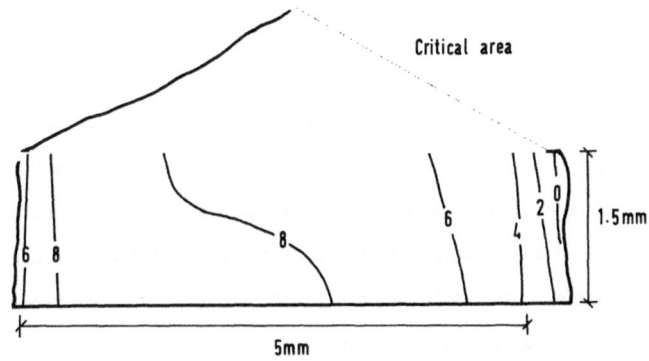

Figure 2. Design 4. Transverse stresses in the composite adherend (MPa) for a 1 MNm^{-1} applied load

In comparison with design 1, the stress concentration at the corner has been avoided and the stress variation through the thickness of the CFRP is now small. A fillet angle of just under 35° reduces the maximum transverse stress in the CFRP to only one-third of that of the basic design. The angle of the fillet also influences the position at which the maximum stress occurs. For fillet angles less than 35°, $\hat{\sigma}_T$ is roughly at point B in Fig 1; this is inside the adhesive fillet but approximately 0.5 mm outside the overlap. The position of B is relatively insensitive to the fillet angle, as for fillet angles less than 35°, the value and location of the maximum stress varies little.

A combination of an (internally) tapered steel adherend with an adhesive fillet results in further reductions in the transverse stress concentration (Design 5). In effect, the transverse stiffness is reduced at the edge of the overlap and, with the addition of an adhesive fillet, the $\hat{\sigma}_T$ stresses are now reduced to about a sixth of those in Design 1.

202

Adhesive Stresses

The finite element analyses give the values of the stress components within the adhesive. From these values, the principal stresses (and hence strains) can be determined in both magnitude and direction. Cohesive failure of the adhesive occurs in regions of maximum stress or strain concentration and results in cracks which run at right angles to the direction of these (stress or strain) maxima. The principal stress distributions therefore indicate the likely locations and directions of failure in the adhesive. Not only can the joint strength then be predicted, but the fracture surfaces can be interpreted.

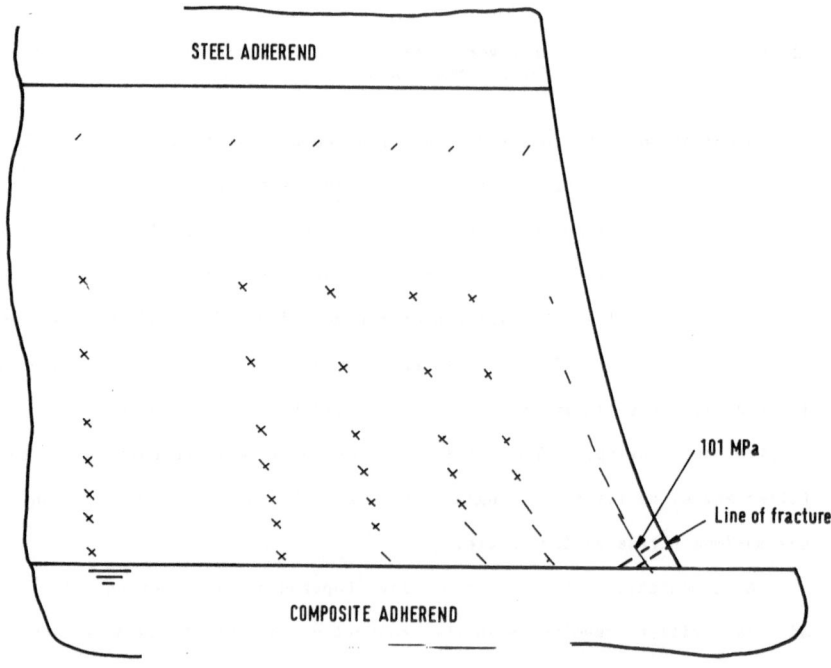

Figure 3. Principal stress distribution (elastic adhesive) at the edge of the adhesive layer for design 1, for a 1 MNm^{-1} applied load.

In design 1 in Fig 3, the highest adhesive principal stress is near to the interface with the central CFRP adherend. The direction of the stress is such that any crack initiating in this region will run towards the interface. In the case of design 4, failure is expected to initiate near the corner of the outer steel adherend. The resulting crack will propagate both through the fillet to the free surface and in the other direction to the interface with the CFRP. With the internally tapered steel adherend, removing the corner relieves the stress concentration so that the maximum tensile stress occurs near the free surface of the adhesive fillet and adjacent to the outer steel adherend corner. Again, cracks initiated in this region would be expected to propagate through the fillet to the interface with the CFRP as indicated.

CONCLUSIONS

There are two possible mechanisms of failure for the double lap joints considered here. In one case, transverse tensile stresses at the edge of the joint close to the interface result in interlaminar failure of the CFRP. In the other case, concentrations of the principal stresses in the adhesive result in tensile (cohesive) failure. The cohesive failure results in cracks which run through the adhesive to the interface, after which the CFRP will fail transversely in an interlaminar manner. However, it may not be clear in the first instance which mechanism is responsible for failure from the fractured surfaces of the joint.

By using suitable failure criteria with the finite element results for the adhesive and the adherends, the load required for failure by either mechanism can be predicted. For interlaminar failure of the particular composite used here, a maximum tensile transverse stress of 40 $\pm$ 6 MPa has been found experimentally. Cohesive failure of the adhesive was predicted by using a maximum principal tensile strain which is the strain at failure in bulk uniaxial tension, which is very similar to the state of stress in the critical regions of the adhesive.

Thus, by using finite element techniques, it is possible not only to predict the strength of joints from fundamentals, but also to predict the mode of failure. This greatly assists the post-failure analysis of joints as it otherwise is difficult, if not impossible, to decide where the failure initiated.

REFERENCES

1. Cushman, J.B., McClesky, S.F. and Ward, S.H., Design, Fabrication and Test of Graphite/Polyimide Composite Joints and Attachments. NASA CR3601. 1983.

2. Volkersen, O., Luftfahrtforschung, 1938, 15, 41.

3. Goland, M. and Reissner, E., J. Appl. Mechs., Trans ASME, 66, A17.

4. Thamm, F., J. Adhesion. 1981, 13, 301.

5. Crocombe, A.D. and Adams, R.D. J. Adhesion, 1981, 13, 141.

6. Volkersen, O. Construction Metallique, 1965, 4, 3.

7. Adams, R.D. and Wake, W.C., Structural Adhesive Joints in Engineering Applied Science Pub., 1984.

8. Adams, R.D. and Peppiatt, N.A. J. Strain Anal., 1974, 9, 185.

9. Adams, R.D., Coppendale, J. and Peppiatt, N.A., Adhesion-2, K.W. Allen, Applied Science Pub., London, 1978, 105.

10. Harris, J.A. and Adams, R.D., Int. J. Adhesion & Adhesives, 1984, 4, 64.

11. Adams, R.D. and Harris, J.A., Int. J. Adhesion and Adhesives, 1987, 7, 69.

12. Raghava, R., Caddell, R.M. and Yeh, G.S.Y., J. Mater. Sci., 1973, 8, 225.

STICK-SLIP AND PEELING
OF ADHESIVE TAPES

Daniel MAUGIS and Michel BARQUINS
Laboratoire de Mécanique des Surfaces du CNRS
LCPC, 58, Bd Lefebvre
75732 Paris Cédex 15

INTRODUCTION

The everyday experience shows that in a given range of temperatures and velocities, the peeling of an adhesive tape is jerky with emission of a characteristic noise. This phenomenon of self sustained oscillations (stick-slip) often described for peeling [1-10] is also observed in other fracture mechanics geometries such as tearing [11-13], wedge loaded double cantilever beams (DCB) [14-22], double torsion [23-29] and blister pressurized by incompressible liquid [30]. All these tests are characterized by the fact that the strain energy release rate G does not increase with the crack length, otherwise a single velocity jumps would occur [31]. The main experimental results are the following :

1/ When the imposed cross-head velocity increases, stick-slip appears abruptly with a large amplitude (defined by the difference between G_i for crack initiation and G_a for crack arrest) at a given velocity (depending on temperature), and its amplitude decreases as the cross-head velocity is further increased, until stable propagation is observed anew [1,7,24,26,27]. Generally it is the peak value G_i which decreases, G_a remaining more or less constant or increasing slowly.

2/ With some materials two stick-slip regions can be observed in the G(v) curve [4-6,17].

3/ The wave length of the stick-slip (the length of each crack jump) increases with the crack length or the peeled length [6] reflecting the increase in the specimen compliance, as observed by inserting a plastic wedge instead of an aluminium wedge in a DCB [21], or by inserting a soft spring between the testing machine and the peeled tape [6].

4/ The shape of the recorded force versus time can be a saw tooth, a sinusoid, or chaotic [8].

Velocity jump and stick-slip were early associated with a decreasing resistance to crack propagation above a given velocity [2,11] and relaxation oscillations due to jumps between two branches of the G(v) curve has been proposed [15,32-34].

In order to understand this complex phenomenon an apparatus was built and preliminary experimental and theoretical results (briefly given in [35,36] are presented hereafter.

APPARATUS

The apparatus (fig. 1) is similar to that used by Satas [8] and consists into an adhesive roller tape (Scotch 3M crystal 602) of radius R unwound at a given linear velocity V (up to 20 m/s) by a couplemeter motor allowing the peel force P to be measured. In a modified version the winding roller is mounted on an elastic plate, the deflection of which giving directly the peel force. The length L of the peeled band can be chosen (between 0,1m and 3m) by changing the distance between the two rollers. Experiments are performed at 23°C in a room humidity of about 50%. The Young modulus of the polyester back is E = 2150 MPa, its thickness h = 40.5 μm and its width b = 19 mm.

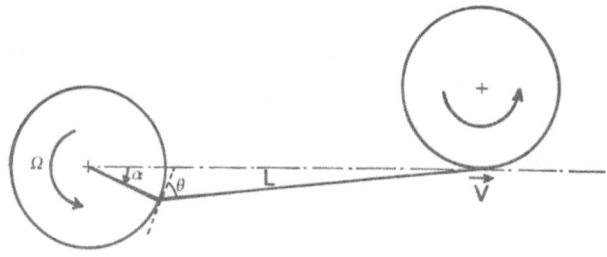

Figure 1. Schematic diagram of peeling geometry.

When peeling is continuous, the peel angle is $\theta = \pi/2$, the energy release rate is :

$$G = P/b = Eh\delta/L \qquad (1)$$

where δ is the peeled band elongation, the crack velocity v (peel velocity) is equal to the linear velocity V imposed by the winding roller, and the angular velocity of the unwound roller is $\Omega = v/R$.

When peeling is jerky, the peeling point oscillates along the circumference of the unwinding roller whose angular velocity is not constant, and the peel angle can vary from 0 to π.

EXPERIMENTAL RESULTS

Figure 2 displays the G(v) curve so obtained, with its two branches of stable propagation separated by a region where stick-slip occurs. The first branch can be represented by the equation :

$$G = w [1 + \alpha(T)v^{n_1}] \qquad (2)$$

where $w = 3J/m^2$ and $n_1 = 0.35$. The parameter $\alpha(T)$ is equal to 135 SI units at 23°C. This branch ends up at point A where $v_c = 6.12 \cdot 10^{-2}$ m/s and $G_c = 170$ J/m^2. At point E ($v = 10^{-3}$ m/s) a transition from cohesive to adhesive rupture, with no change in the slope of the G(v) curve is noted, in agreement with the observations of Aubrey and Scheriff [7]. The second branch with a positive slope begins at point C ($v_2 = 2.1$ m/s, $G_{min} = 30$ J/m^2) with a near zero slope until point F ($v_4 = 10$ m/s) beyond which the curve is descrived by the equation

$$G = G_c(v/v_1)^{n_2}$$

with $v_1 = 18$ m/s (point B) and $n_2 = 5.5$.

The signal delivered by the couplemeter was found somewhat noisy and was filtered before measuring the frequency of oscillations as a function of peeled length L and imposed velocity V. However these frequencies were in agreement with the marks left on the tape at each cohesive-adhesive transition (these marks disappear for V > 0.65 m/s). Figures 3 and 4 display these results and clearly show that the frequency increases with the velocity V and decreases with the peeled length L.

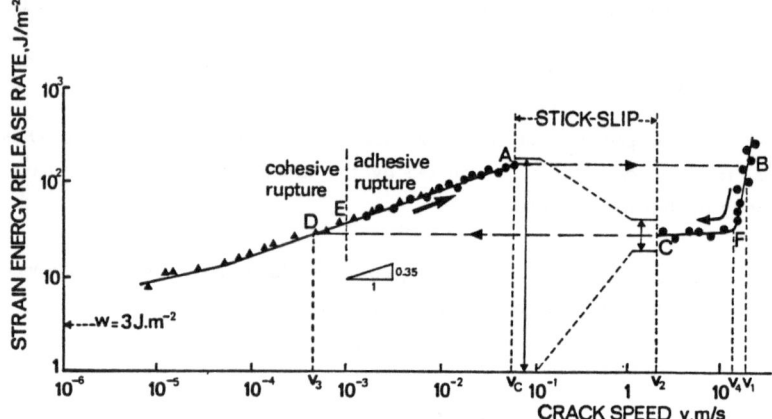

Figure 2. Experimental G(v) curve.

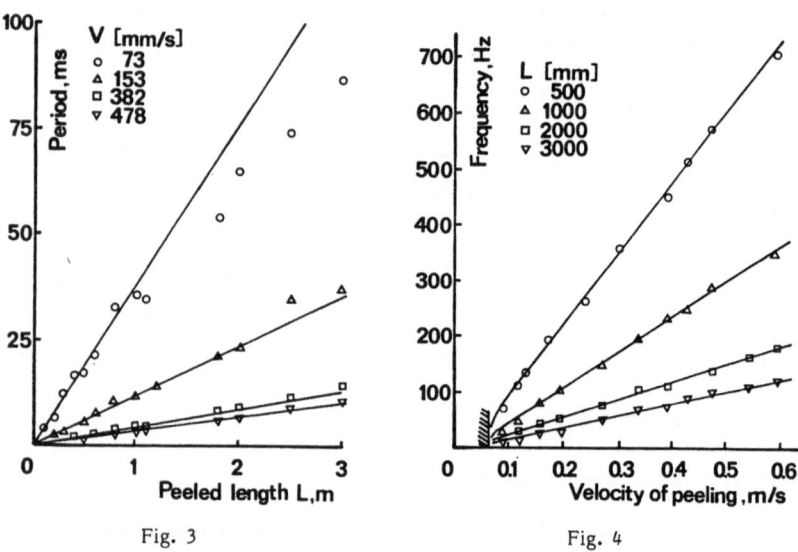

Fig. 3   Fig. 4

Figure 3. Period of oscillations versus peeled length for various imposed velocities.

Figure 4. Frequency of oscillations versus imposed velocity for various peeled
lengths.

THE RELAXATION OSCILLATION MODEL

Due to the negative resistance to crack propagation between v_c and v_2, a stable crack propagation cannot be observed for an imposed velocity V in this range. Instead, the first branch is followed up to G_c (point A) where the velocity v_2 is too low, so that the velocity jumps to v_1 on the second branch with positive resistance (point B) where the velocity is too high. The crack slows down to point C where the velocity v_2 is still too high, then jumps to point D where the crack seems arrested (*). Then, the velocity increases to v_c (too low) and jumps to B. A stick-slip motion thus occurs, where most of the time is spent on the slow velocity branch DA. At each cycle a cohesive-adhesive transition occurs, with a mark left on the tape.

The period of oscillations is thus the sum of times spent on the branches DA and BC, the second being negligible compared to the first. Neglecting the variation of peel angle around $\theta = \pi/2$ and the corresponding variation in peeled length, the relation between v and V is :

$$v = V - d\delta/dt \qquad (3)$$

Extracting v from equ.(2), putting it in equ.(3) and using equ.(1), the time spent on DA, and thus the period is :

$$t = (L/Eh) \int_{G_{min}}^{G_c} dG/(V-v_c((G-w)/(G_c-w))^{1/n_1}) \qquad (4)$$

This period is proportionnal to the peeled length L. Numerical calculation is shown as full line in figures 3 and 4 and is in satisfactory agreement with experimental results.

The model thus accounts for the frequencies observed in this range, for the acoustic emission at each velocity jump, for the marks left on the tape, but does not explain why the amplitude of oscillations decrease with V, why the oscillations are sinusoidal instead of saw-tooth shaped, and why marks disappear at V> 0.65 m/s.

(*) Incidentally such an hysteresis between accelerating cracks and decelerating crack has been observed by Kobayashi and Dally [37,38] on epoxy resins.

SIMPLE MODEL FOR INERTIAL EFFECTS

In presence of stick-slip, inertial effects cannot be neglected, and the law of energy conservation leads to :

$$G = \phi(v) - dU_k/dA \qquad (5)$$

where :

$$\phi(v) = w + w\varphi(v) \qquad (6)$$

represents the relation between G and v for stationary crack velocities, with the negative resistance branch (*) between the two positive resistance branches. $U_K = I\Omega^2/2$ is the kinetic energy of the system (i.e. of the unwinding roller with inertia I and angular velocity Ω) and dA = - bvdt is the variation of contact area.

We will neglect again the variation of peel angle during stick-slip so that the angular velocity is :

$$\Omega = v/R \qquad (7)$$

and we will neglect also the change in inertia and radius during unwinding. Note that the model cannot be exact, since the equation $v = \Omega R$ assume vanishing inertia in contradiction with the hypothesis. Nevertheless, the structure of resulting equations is very interesting.

Letting $m = I/R^2$, equ.(6) and (7) leads to :

$$\dot{v} = dv/dt = b[G - \phi(v)]/m \qquad (8)$$

which shows that the curve G(v) must be carefully distinguished from the $\phi(v)$ curve when the crack velocity is not constant.

By time derivation of the force-elongation relation $P = k\delta$ where $k = Ebh/L$ is the stiffness of the peeled length, and using equ.(3) one has :

(*) The shape of this branch corresponding to unstable stationary equilibrium cannot be determined directly.

$$\overset{\bullet}{G} = dG/dt = - k(v-V)/b \tag{9}$$

By derivation of equ.(8) and letting $\omega = \sqrt{k/m}$ and $\mu = b/\sqrt{km}$ one has :

$$\overset{\bullet\bullet}{v} + \rho\omega(d\phi/dv)\overset{\bullet}{v} + \omega^2(v-V) = 0 \tag{10}$$

Let us take the point $[V, \phi(V)]$ (stable or unstable) as the origin, (Fig. 5) and let :

$$x = v - V$$

$$F(x) = \phi(v) - \phi(V)$$

$$f(x) = dF/dx = d\phi/dv$$

one has

$$\overset{\bullet\bullet}{x} + \mu\omega f(x) \overset{\bullet}{x} + \omega^2 x = 0 \tag{11}$$

which is the classical Lienard equation for self- sustained oscillations, and which is known to have limit cycles when the function $F(x)$ has a branch with a negative slope [39,40].

Letting :

$$y = G - \phi(v)$$

this equation is equivalent to the autonomous system :

$$\overset{\bullet}{x} = \mu\omega[y - F(x)] = X(x,y) \tag{12}$$

$$\overset{\bullet}{y} = - (\omega/\mu)x = Y(x,y) \tag{13}$$

which is a flow in $\mathbb{R}^2$ describing the motion in the space phase, or Lienard plane (*), (x,y), i.e. (v,G) with a translation.

(*) In fact the classical Lienard tranformation would be :

$$\overset{\bullet}{x} = y_1 - \mu\omega F(x)$$

$$\overset{\bullet}{y}_1 = -\omega^2 x$$

One has put here $y = y_1/\mu\omega = (b/m)y_1$ in order to have in the plane (x,y) a simple representation of (v,G) and (v, ϕ). Circular orbits in the classical Lienard plane would give ellipses here.

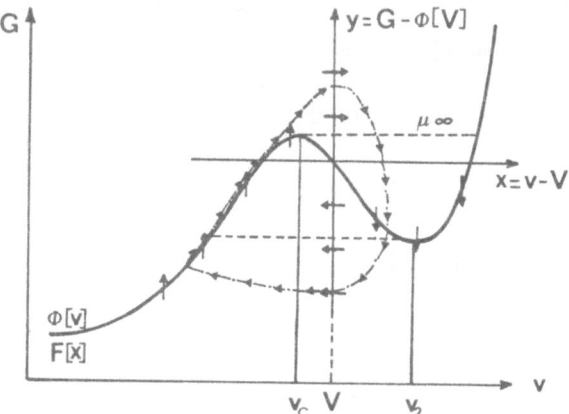

Figure 5. Diagram of limit cycles on the G(v) curve.
Relaxation oscillations are only observed for
zéro inertia ($\mu\,\infty$).

The energy of the system is :

$$E = (m\dot{x}^2 + kx^2)/2$$

and its variation with time is :

$$\dot{E} = \dot{x}\,(m\ddot{x} + kx)$$
$$= - b\dot{x}^2 f(x)$$

Energy increases when $f(x) < 0$, it decreases when $f(x) > 0$, and a limit cycle corresponding to established oscillations is reached when this energy remains constant on a period. A limit cycle cannot be observed in the only range $x_c < x < x_2$ but must cross the limits $x = x_c$ and/or $x = x_2$.

The slope of orbits in this space is :

$$dy/dx = - x/\mu^2[y - F(x)] \qquad (14)$$

and one can see in Figure 5 that the axis $x = 0$ is an isocline $0°$, that the curve $y = F(x)$ (i.e. the curve $G = \phi(v)$) is an isocline $90°$, and that the origin is a singular point. When the inertia term m tends towards zero ($\mu \rightarrow \infty$), equ.(14) becomes :

$$dy[y - F(x)] = 0$$

which represents the curve $y = F(x)$ and the horizontal $dy = 0$ on which, from equ.(8) the velocity is infinite. It is thus only for $m \rightarrow 0$ that the relaxation cycle previously described is valid.

Linearization of the system : behavior at the origin

Let us linearize the system of equations (12,13) around the origin $x = y = 0$ (i.e. $v = V$, $dv/dt = 0$) by writing :

$$\dot{x} = X(x,y) = \partial X/\partial x \ (0,0)x + \partial X/\partial y \ (0,0)y = Ax + By$$

$$\dot{y} = Y(x,y) = \partial Y/\partial x \ (0,0)x + \partial Y/\partial y \ (0,0)y = Cx + Dy$$

Letting : $$2S = \mu \ (d\phi/dv)_{v \ =V}$$

it comes :

$$\begin{pmatrix} \dot{x} \\ \dot{y} \end{pmatrix} = \omega \begin{pmatrix} -2S & \mu \\ -1/\mu & 0 \end{pmatrix} \begin{pmatrix} x \\ y \end{pmatrix} \tag{15}$$

The eigenvalues of the system are solutions of the characteristic equation :

$$\lambda^2 - \lambda(A + D) + (AD - BC) = 0$$

i.e. $$\lambda^2 + 2S\omega\lambda + \omega^2 = 0$$

whose discriminant is :

$$\Delta = 4\omega^2(S^2 - 1)$$

The solutions are well known [39,40] :

1/ - When $|S| > 1$, there are two roots of same sign :

$$\lambda_1, \lambda_2 = - S\omega \pm \omega\sqrt{S^2 - 1}$$

positive if $S < 0$ and negative if $S > 0$. The origin is a node (unstable if $S < 0$, stable if $S > 0$) and the orbits respectively leave or approach the origin in an aperiodic motion. The two eigenvectors are :

$$x = \mu\omega(S + \sqrt{S^2 - 1})$$

$$y = 1$$

and

$$x = \mu\omega(S - \sqrt{S^2-1})$$
$$y = 1$$

All orbits except two are tangent to an eigenvector and have the second as parabolic asymptot.

2/ When $0 < |S| < 1$ there are two complex roots :

$$\lambda_1, \lambda_2 = -S\omega \pm i\omega'$$
$$\text{with } \omega' = \omega\sqrt{1-S^2}$$

and the origin is a focus. The orbits are spirals expanding from the origin if $S < 0$ or approaching it if $S > 0$.

3/ When $S = 0$, the roots are pure imaginary :

$$\lambda_1, \lambda_2 = \pm i\omega$$

and the origin is a centre. It is a stable equilibrium point (but with marginal stability) and the orbits are spirals around the origin, as shown by equation

$$dy/dx = (Cx+Dy)/(Ax+By) = -(1/\mu^2)(x/y)$$

i.e. $\qquad\qquad y^2 + x^2/\mu^2 = Cte.$

Oscillations are sinusoidal with the angular frequency ω.

Hopf bifurcations around the extremums of $\phi(v)$

Let us assume that the $\phi(v)$ curve is a smooth function. One has $d\phi/dv = 0$, i.e. $S = 0$, at the maximum (G_c) and the minimum G_{min}. These two points $(V = v_c$ and $V = v_2)$ are centres corresponding to sinusoïdal oscillation of angular frequency $\omega = \sqrt{k/m}$. In their neighbouring, the origins (stable or unstable) are foci and the orbits are spirals (S is small). When $S < 0$ these spirals expand towards a limit cycle (stable limit cycle), but when $S > 0$ the contracting spirals, sometimes, can come also from a limit cycle (unstable limit cycle). Let us study this point.

Let us return to equ.(11) and take $\tau = \omega t$ as unit time. It comes :

$$d^2x/d\tau^2 + \mu f(x)dx/d\tau + x = 0 \qquad\qquad (16)$$

which is equivalent to :

$$dx/d\tau = y_1 - \mu F(x)$$
$$dy_1/d\tau = - x$$

or to :

$$xdx + y_1\,dy_1 + \mu F(x)dy_1 = 0 \qquad (17)$$

The integral of $xdx + y_1 dy_1$ being zero on a closed path from the Riemann theorem, one has on a closed path (limit cycle) :

$$\oint F(x)dy_1 = 0 \qquad (18)$$

which is the Lienard condition of a limit cycle [39].

Let us develop $F(x)$ around $x = 0$:

$$F(x) = F(0) + xdF/dx(0) + (x^2/2)d^2F/dx^2(0) + (x^3/6)\ d^3F/dx^3(0)$$

$$= x(d\phi/dv)_{v=V} + (x^2/2)\ (d^2\phi/dv^2)_{v=V} + (x^3/6)\ (d^3\phi/dv^3)_{v=V}$$

and place the origin $x = 0$ in the neighbouring of an extremum where in this phase space (x,y_1) the orbits are circular (the linearized equation gives $x^2 + y_1^2 = Cte$). The limit circle will be approximately circular, and we can write :

$$x = R \sin \theta$$
$$y = R \cos \theta$$

In the integral (18) the even terms disappear, and the Lienard condition gives :

$$R = \sqrt{- 8F'(0)/F'''(0)} \qquad (19)$$

If the third derivative of $\phi(v)$ is positive near an extremum, limit cycles are observed only around origins situated on the branch with a negative slope (Fig. 6) (unstable equilibrium points), and their amplitude increases as the square root of the bifurcation parameter - S. This bifurcation from stable solutions $G = \phi(v)$ to stable limit cycles of increasing size when $(d\phi/dv)_{v=V}$ becomes negative is a supercritical Hopf bifurcation.

If the third derivative of $\phi(v)$ is negative, it is stable equilibrium point ($S>0$) which are surrounded by an unstable limit cycle (inverted Hopf bifurcation) itself surrounded by a stable limit cycle, as would show the stroboscopic equation deduced from the development of F(x) at the fifth order [39]. The extremum is thus a <u>subcritical Hopf bifurcation</u> (Fig. 7). Oscillations appear abruptly at S = 0 when S decreases and disappear abruptly at the coalescence of the two limit cycles when S increases (hysteresis).

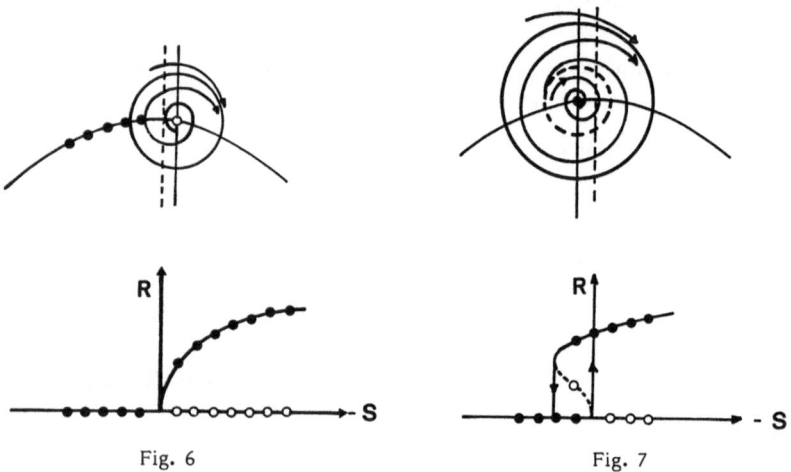

Fig. 6   Fig. 7

Figure 6. Supercritical Hopf bifurcation near an extremum of the ϕ(v) curve. An unstable point is surrounded by a stable limit cycle.

Figure 7. Supercritical Hopf bifurcation near an extremum of the ϕ(v) curve. Stable equilibrium points are surrounded by an unstable limit cycle surrounded itself by a stable limit cycle.

For a viscoelastic solid, the dissipation term near the maximum can be written as :

$$\phi(v) = w \, [1 + Av^n/(1+Bv^m)]$$

with $0 < n < 1$, and a tedious calculation shows that the third derivative at $v = v_c$ is :

$$(d^3\phi/dv^3)_{v \, = \, v_c} = wAv^{n-3} \, n(m-n)^2 \, (2n - m +3)/m$$

For a symetrical peak in a $(\log \phi, \log v)$ diagram, i.e. $m = 2n$, its value is positive, so that oscillations will appear progressively, when V will exceed the critical velocity v_c corresponding to this maximum.

CONCLUSION

This simple model introduces to the classical field of non-linear oscillations. It explains why the oscillations can be sinusoidal or saw-tooth shaped depending on the shape of the limit cycle, i.e. on the inertia of the system, and why the amplitude of oscillations changes with the imposed velocity V. It shows that G_i for crack initiation and G_a for crack arrest (the upper and lower points of the $G(v)$ curve) are different from G_c and G_{min} (excepted for the relaxation cycle) and are not a material property as suspected by some investigators [41,42]. The fact that limit cycles closely follow the first positive branch explains the relative success of the calculation of frequencies based on the time spent on the first positive branch.

However the reality is more complex since complicated and even chaotic signals have been observed as described below.

FURTHER INVESTIGATIONS

As noted above it was found that the signal delivered by the couplemeter was found noisy and was filtered. It was not clear if the noise arose from the couplemeter itself or was a true effect. In a modified version of the apparatus, the winding roller was mounted or an elastic leaf of spring whose deflection directly gives the peeling force.

It is found that, at the end of the subcritical crack propagation $(V = v_c)$, a slow increase in the peeling velocity provokes the appearance of sporadic unstabilities (Fig. 8a) which become more and more frequent and regular (Fig. 8b) and then periodical (Fig. 8c,d,e) with a slow and continuous increase in the amplitude up to V = 30 cm/s. Then, a chaotic behaviour occurs with a marked decrease in the maximum value of the peeling force (Fig. 8f,g,h) until the peeling speed $V = v_2$ is reached. Recordings shown in Fig. 8a-h clearly prove that oscillations are closely related to the eigenfrequencies of the elastic plate supporting the winding roller.

It is not clear at this stage if this complex behaviour is due to coupled oscillations between the elastic plate and the rotation of the unwinding roller. Nevertheless, closer examination of the dynamical system using the couplemeter, i.e. without coupled oscillations, shows that the behaviour could be complex. In

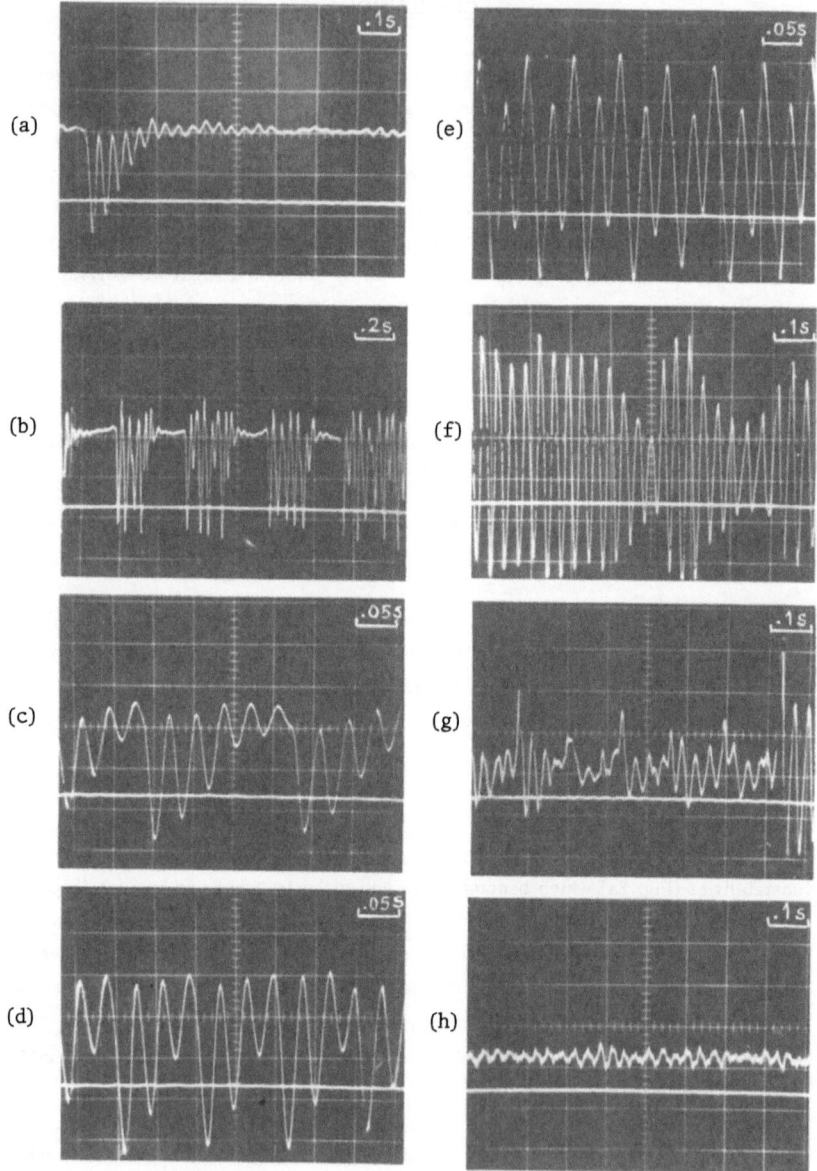

Figure 8. Recordings of the peeling force at increasing imposed velocity.

fact, there are three parameters of control at the disposal of the experimenter : the imposed velocity V, the distance ℓ between the axis of the rollers, and the inertia I of the unwinding roller. There are seven variables : the stretching force in the peeled tape (the peeling force P), its elongation δ , the peel angle θ , the peeling point on the unwinding roller, the geometrical length L of the peeled tape, the crack velocity v along the unwinding roller, and the angular velocity of the unwinding roller. Seven equation can be derived, which reduce to three differential equation of the first order, one of which being non linear (due to the non-linear damping ϕ (v)). It is well known (see for example ref.[43,44]) that in such systems complex behaviours can be observed with frequency doubling, intermittencies and chaotic motions when limit cycles are changed into strange attractors. Work is in progress in this direction.

ACKNOWLEDGMENTS

The authors are indebted to DRET and CEA for financial support.

REFERENCES

1. Gardon, J.L., Pell adhesion I. Some phenomenological aspects of the test. J. Appl. Polym. Sci., 1963, 7, 625-41.

2. Kaelble, D.H. Theory and analysis of peel adhesion : rate-temperature dependence of viscoelastic interlayers. J. Colloid Sci., 1964, 19, 413-24.

3. Aubrey, D.W., Welding, G.N. and Wong, T., Failure mechanisms in peeling of pressure-seensitive adhesive tape. J. Appl. Polym. Sci., 1969, 13, 2193-207.

4. Gent, A.N. and Petrich, R.P., Adhesion of viscoelastic materials to rigid substrates. Proc. Roy. Soc., 1969, A310, 433-48.

5. Aubrey, D.W., Pressure-sensitive adhesives. Principles of formulation. In Developments in Adhesives - I, ed. W.C. Wake, Appl. Sci. Publ., London, 1977, pp. 127-56.

6 Aubrey, D.W., Viscoelastic basis of peel adhesion. In Adhesion 3, ed. K.W. Allen, Appl. Sci. Publ., London, 1978, pp. 191-205.

7 Aubrey, D.W. and Scherriff, M., Peel adhesion and viscoelasticity of rubber-resin blends. J. Polym. Sci. Polym. Chem. Ed., 1980, 18, 2597-608.

8 Satas, D., Peel. In Handbook of Pressure-sensitive adhesive technology, ed. D. Satas, Van Nostrand, New York, 1982, pp. 50-77.

9. Hamed, G.R. and Shieh, C.H., Relationship between the cohesive strength and the tack of elastomers. J. Polym. Sci. Polym. Phys. Ed., 1983, 21, 1415-25.

10. Aubrey, D.W., Effect of adhesive composition on the peeling behaviour of adhesive tapes. In Adhesion 8, ed. K.W. Allen, Elsevier Appl. Sci. Publ., London, 1984, pp. 19-32.

11. Greensmith, H.W. and Thomas, A.G., Rupture of rubber III. Determination of tear properties. J. Polym. Sci., 1955, 18, 189-200.

12. Veith, A.G., A new tear test for rubber. Rubber Chem. Technol., 1965, 38, 700-18.

13. Stracer, R.G., Yanyo, L.C. and Kelley, F.N., Observations on the tearing of elastomers. Rubber Chem. Technol., 1985, 58, 421-35.

14. Broutman, L.J. and Mc Garry, F.J., Fracture surface work measurements on glassy polymers by a cleavage technique. I. Effect of temperature. J. Appl. Polym. Sci., 1965, 9, 589-608.

15. Clark, A.B.J. and Irwin, G.R., Crack propagation behaviors. Exp. Mech., 1966, 6, 321-30.

16. Ripling, E.J., Mostovoy, S. and Corten, H.T., Fracture mechanics : a tool for evaluating structural adhesives. J. Adhesion, 1971, 3, 107-23.

17. Broutman, L.J. and Kobayashi, T., Dynamic crack propagation studies. In Dynamic crack propagation, ed. G.C. Sih, Noordhoff, Leyden, 1973 pp. 215-25.

18. Rosenfield, A.R. and Kanninen, M.F., The fracture mechanics of glassy polymers. J. Macromol. Sci. Phys., 1973, B7, 609-31.

19. Selby, K. and Miller, L.E., Fracture toughness and mechanical behaviour of an epoxy resin. J. Mater. Sci., 1975, 10, 12-24.

20. Mostovoy, S., Crosley, P.B. and Ripling, E.J., Note on the dependence of crack velocity on driving force for an epoxy resin. In Cracks and fracture, ASTM STP 601, Philadephia, 1976, pp. 234-44.

21. Fourney, W.L. and Kobayashi, T., Influence of loading system on crack propagation and arrest behavior in a double-cantilever beam specimen. In Fracture mechanics applied to brittle materials, ASTM STP 678, Philadelphia, 1979, pp. 47-59.

22. Takahashi, K. and Mada, T., Ultrasonic fractography studies on disconti-nuous fracture propagation in polymers. Jap. J. Appl. Phys., 1985, 24, 196-8.

23. Young, R.J. and Beaumont, P.W.R., Crack propagation and arrest in epoxy resins. J. Mater. Sci., 1976, 11, 776-9.

24. Yamini, S. and Young, R.J., Stability of crack propagation in epoxy resins. Polymer, 1977, 18, 1075-80.

25. Phillips, D.C., Scott, J.M. and Jones, M., Crack propagation in an amine-cured epoxide resin. J. Mater. Sci., 1978, 13, 311-22.

221

26. Gledhill, R.A., Kinloch, A.J., Yamini, S. and Young, R.J., Relationship between mechanical properties of and crack propagation in epoxy resin adhesives. Polymer, 1978, 19, 574-82.

27. Yamini, S. and Young, R.J., Crack propagation in and fractography of epoxy resins. J. Mater. Sci., 1979, 14, 1609-18.

28. Yamini, S. and Young, R.J., The mechanical properties of epoxy resins. Part 2. Effect of plastic deformation upon crack propagation. J. Mater. Sci., 1980, 15, 1823-31.

29. Spanoudakis, J. and Young, R.J., Crack propagation in a glass particle-filled epoxy resin. Part 1. Effect of particle volume fraction and size. J. Mater. Sci., 1980, 19, 473-86.

30. Dannenberg, H., Measurement of adhesion by a blister method. J. Appl. Polym. Sci., 1961, 5, 125-34.

31. Maugis, D., Subcritical crack-growth, surface energy, fracture toughness, stick-slip and embrittlement. J. Mater. Sci., 1985, 20, 3041-73.

32. Williams, J.G., Radon, J.C. and Turner, C.E., Designing against fracture in brittle plastics. Polym. Eng. Sci., 1968, 4, 130-41.

33. Maugis, D., Adherence of solids. In Microscopic aspects of adhesion and lubrication, ed. J.M. Georges, Elsevier, Amsterdam, 1982, pp. 221-52.

34. Williams, J.G., Fracture mechanics of polymers. Ellis Horwood, 1984.

35. Barquins, M., Khandani, B. and Maugis, D., Propagation saccadée de fis-sure dans le pelage d'un solide viscoelastique. Comptes Rend. Ac. Sci. Paris, série II, 1986, 303, 1517-19.

36. Maugis, D., Propagation saccadée de fissure en pelage, rôle de l'inertie. Comptes Rend. Ac. Sci. Paris, (in press).

37. Kobayashi, T. and Dally, J.W., Relation between crack velocity and the stress intensity factor in birefringent polymers. In Fast fracture and crack arrest, ASTM STP 627, Philadelphia, 1977, pp. 257-73.

38. Kobayashi, T. and Dally, J.W., A system of modified epoxies for dynamic photoelastic studies. Exper. Mech., 1977, 17, 367-74.

39. Minorsky, N., Non-linear oscillations, Van Nostrand, New York, 1962.

40. Jordan, D.W. and Smith, P., Non linear ordinary differential equations. Clarendon Press, Oxford, 1977.

41. Dally, J.W., Fourney, W.L. and Irwin, G.R., On the uniqueness of the stress intensity factor-crack velocity relationship. Int. J. Fracture, 1985, 27, 159-68.

42. Kobayashi, A.S., Ramulu, M., Dadkhah, M.S. Yang, K.H. and Kang, B.S.G., Dynamic fracture toughness. Int. J. Fracture, 1986, 30, 275-285.

43. Bergé, P., Pomeau, Y. and Vidal, Ch., L'ordre dans le chaos. Hermann, Paris, 1984.

44. Guckenheimer, J. and Holmes, P., Non linear oscillations, dynamical systems and bifurcations of vector fields. Springer Verlag, New York, 2nd printing, 1986.

HARD RUBBER/METAL ADHESION ASSESSMENT USING A HEAVY CYLINDER ROLLING TEST

M.E.R. Shanahan, N. Zaghzi, J. Schultz.
Centre de Recherches sur la Physico-Chimie des Surfaces Solides

and

Ecole Nationale Supérieure de Chimie de Mulhouse,
3, Rue Alfred Werner,
68093 Mulhouse Cedex,
FRANCE

and

A. Carré
Corning Europe Inc.,
77211 Avon Cedex,
FRANCE

1. INTRODUCTION

When an elastomer is put into contact with a rigid substrate, a certain adhesion, or adherence may develop at the interface. This effect is beneficial under some circumstances, such as in the case of a vehicle tyre in contact with a road surface. In other cases, adherence can cause problems. For example, when a water tap has been left closed for too long, the adhesional force of the rubber washer to metallic components can be superior to the cohesional strength of the elastomeric material and subsequent opening of the tap leads to the washer tearing. A high degree of adhesion may be apparent even at room temperature if the contact time and pressure are sufficient. Nevertheless, a necessary condition for long term adhesion is good initial or spontaneous contact. The present study considers this spontaneous adhesion between a steel substrate and two types of elastomer, both containing carbon-black. A heavy cylinder rolling test has been developed. Cylinder rolling tests have previously been used to study soft rubbers [1]. However, the elastomers employed in the present

context are relatively hard and a high contact pressure is therefore
necessary in order to ensure intimate steel-elastomer contact. This leads
to the fact that the energy dissipated during rolling is related not only
to the hysteretic effects accompanying adherend separation but also to
losses invoked by the stress field in the elastomer beneath the heavy
cylinder.

2. THEORY

Consider a cylinder of mass m, radius R and length L rolling down an
elastomeric track inclined at an angle θ to the horizontal (Fig. 1). After

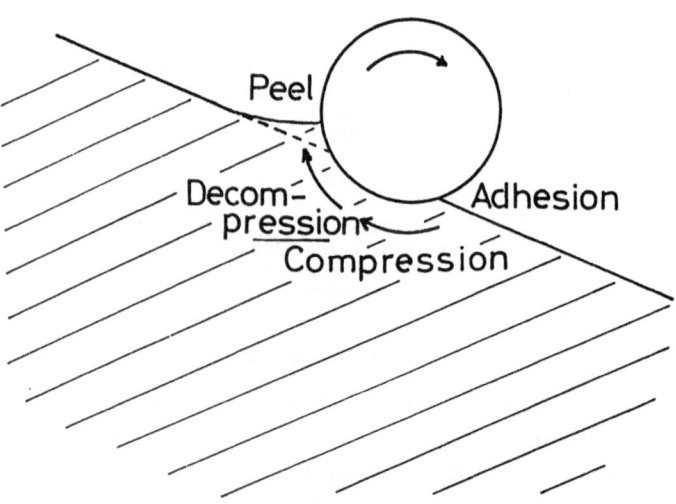

Figure 1   Schematic representation of heavy cylinder
            rolling down elastomeric track.

rolling unit distance, the cylinder will lose potential energy U_p per
unit length [1] g being gravitational acceleration. Were the system to

$$U_p = mg \, L^{-1} \sin \theta \qquad \cdot \quad \cdot \quad \cdot \qquad 1$$

be without mechanical loss, this energy would be transformed into kinetic
energy and the cylinder would accelerate (see below). However, the essential of this potential energy is dissipated by hysteretic losses in the
elastomer and the cylinder maintains a steady speed of descent, V. For a
lightweight cylinder, dissipation is due to the difference in energy
required for the formation and the separation processes involved at the
elastomer-cylinder interface [1]. Again considering unit rolling distance
and unit cylinder length, the formation of contact at the leading edge
involves an energy gain of $(-W_o)$ where W_o is Dupré's reversible work
of adhesion and is given by:

$$W_o = \gamma_s + \gamma_E - \gamma_{SE} \qquad \cdot \qquad \cdot \qquad 2$$

γ_S and γ_E representing the surface free energies respectively of the
steel and the elastomer and γ_{SE} their interfacial free energy [2]. At
the trailing edge, an energy loss of W occurs where W represents the peel
energy of elastomer-metal separation. W is much greater than W_o due to
hysteretic losses accompanying the inevitable elastomer deformation during
separation, but it has been shown that the two are directly proportional
[3,4]. The dissipation factor connecting the two involves test rate,
temperature and average elastomeric molecular weight between crosslinks,
M_c [5]. In the case of a light cylinder rolling at constant speed, U_p
of equation 1 is directly balanced by $(W-W_o)$. However, in the case of a
heavy cylinder, the sheer weight of the cylinder causes a non-negligible
stress field in the elastomer which corresponds to elastic stored energy of
ϕ being gained per unit distance rolled and per unit cylinder length. The
value of ϕ has been calculated by Greenwood et al [6] using the analysis
originally proposed by Hertz [7] for static contact. The calculation by
Greenwood et al was intended to apply to horizontal rolling friction. As a
consequence, in the present context modification is necessary to account
for the fact that the effective force perpendicular to the elastomeric
track must be corrected to allow for the tract inclination of θ with
respect to the horizontal. In equation 3, E and ν represent respectively
Young's modulus and Poisson's ratio for the elastomer. Actually, equation
3 is correct only to first approximation. A more rigorous analysis should
allow for local deformation near the elastomer-cylinder interface caused by
forces of adhesion [8].

$$\phi = \frac{4}{3}\left(\frac{mg \cos \theta}{\pi L}\right)^{3/2} \cdot \left(\frac{1 - v^2}{R E}\right)^{1/2} \quad . \quad . \quad . \quad 3$$

Clearly, if the deformed track is purely elastic, the compressional strain energy gained in front of the rolling cylinder will be recuperated completely behind the cylinder on decompression. However, because of the hysteretic nature of the elastomer, a certain fraction will be dissipated. This dissipational energy is then represented by $\alpha \phi$ where $\phi \leqslant \alpha \leqslant 1$. We are now in a position to write the overall energy balance per unit distance rolled and per unit cylinder length:

$$U_p = W - W_0 + \alpha \phi \quad . \quad . \quad . \quad 4$$

Since, as stated above, W is much greater than W_0, we may reasonably neglect the latter. The problem may be considered from the point of view of fracture mechanics, in which case the trailing edge elastomer-cylinder separation can be regarded as crack growth and the leading edge formation is, in a sence, "negative" crack growth. We may then define a strain energy release rate, G [9]:

$$G = W + \alpha \phi \quad . \quad . \quad . \quad 5$$

To summarise, equation 5 states that the loss of cylinder potential energy during rolling can be considered as a strain energy release rate for which the energy losses correspond to elastomeric dissipation both due to local deformation in the zone of elastomer-cylinder separation and to hysteretic behaviour within the stress field caused by the weight of the cylinder.

3.   EXPERIMENTAL

3.1   Materials

The cylinders used in this study were made from steel 100 C6. A high degree of polish on the rolling surfaces was ensured using a high pressure lathe. Diameter and length, L, of the rolling surfaces were both 2.5 cm. The initially hollow cylinders could be fitted with supplementary cylindrically symmetrical masses so that the influence of cylinder weight could be studied without changing rolling dimensions.

Two crosslinked, carbon-black filled elastomers were used for the rolling tracks:

- an ethylene (52%), propylene (44%) norbornene (4%) terpolymer, referred to as EPDM,

- and an acrylonitrile (34%), butadiene (66%) copolymer, referred to as NBR.

Data concerning the elastomers are given in Table 1. The elastomeric tracks of 5.8 mm thickness with very smooth surfaces were produced by hot compression moulding.

TABLE 1

Constituents and properties of the elastomers

Parts by weight	EPDM	NBR
Vistalon 2504 (ESSO)	100	–
Krynac 3435 (POLYSAR)	–	100
Carbon Black (HAF N330)	50	50
Sulphur	0.5	2
Zinc Oxide	5	5
Accelerator CBS	–	1
Accelerator DTMT	3.3	2
Accelerator MBT	1	–
Stearic Acid	–	1
Conditions of crosslinking	30 min at 160°C	60 min at 150°C
Young's Modulus (MPa)	18	8
Glass Transition Temperature (°C)	−37.5	−26
Mean Intercrosslink Molecular weight, M_c (g mol^{-1})	2,400	6,200

3.2   Rolling set-up

The experimental set-up, shown in Fig. 2, is composed of a supported rectangular metallic block, whose angle with respect to the horizontal can be varied and measured by means of a micrometer gauge. A fixation for the

rubber track is attached to the block. At each end, there is an electro-
magnet for retaining the steel cylinder until the desired moment of
release. Rolling times are measured by means of photoelectric cells at
each end recording the passing of the cylinder.

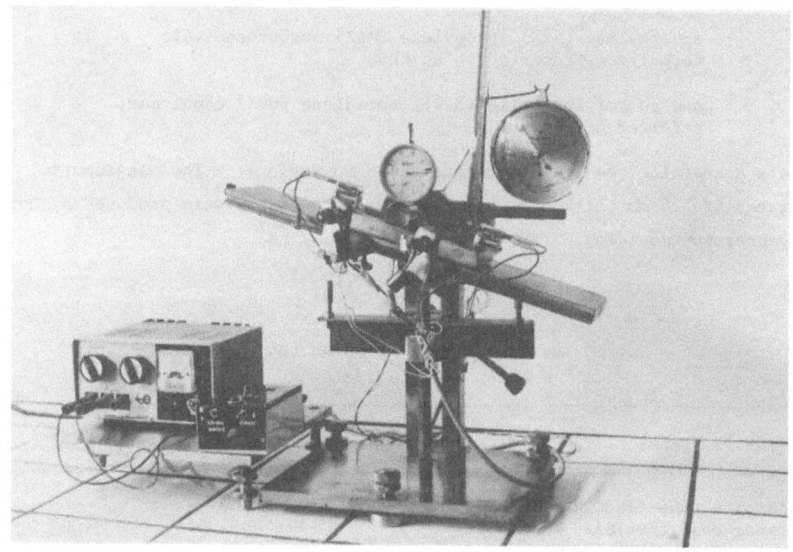

Figure 2   View of experimental set-up
                  (without environmental box).

Preparation of both elastomer and steel surfaces prior to rolling was
limited to washing with ethanol and drying. Experiments were conducted in
an environmental box (not shown in Fig. 2). The tests reported here were
conducted at 20% R.H.

4.   RESULTS

Typical results for rolling experiments conducted both with EPDM and
NBR tracks at 20°C and with various cylinder weights are shown in Figs.
3 and 4 with absiccae representing equilibrium rolling speed, V, and
ordinate, the strain energy release rate, G (equation 5). Both scales are
logarithmic because of the large ranges covered. In both cases, the overall trend is that strain energy release rate increases markedly with
rolling speed. This is in good agreement with what would be expected from

hysteretic energy dissipation in the elastomer. The bulk dissipative term, 0, of equation 5 is expected to obey the classical law of Williams, Landel and Ferry (WLF) [10] involving the time-temperature superposition principle of viscoelastic polymer deformation. In addition, the term W may be expressed as $W_o f$ (V,T) where f (V,T) again is a dissipation factor obeying the WLF principle [3].

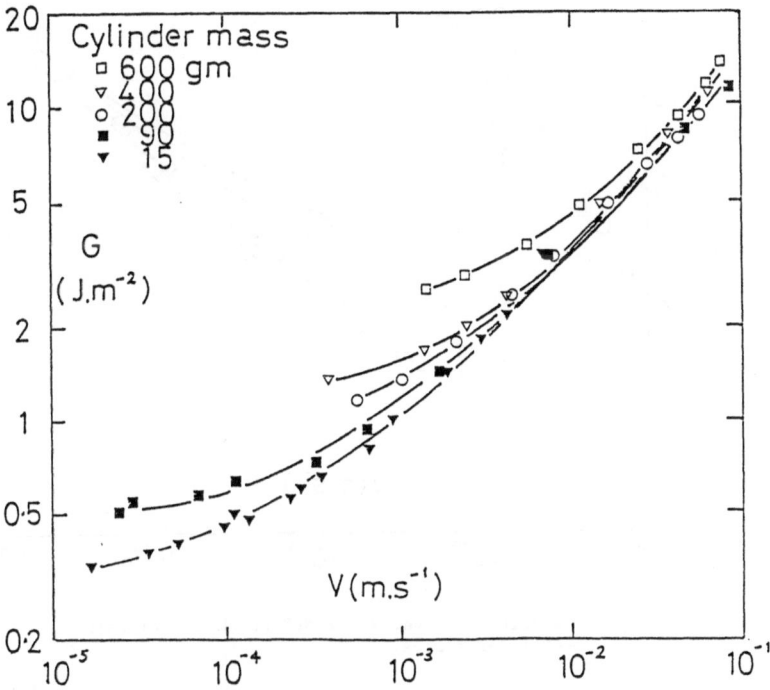

Figure 3   Results of G vs. V for EPDM track and various cylinder masses.

Assuming that both bulk and separation contributions to hysteresis obey the same functional relationships, we should be able to write:

$$G = G_o \psi (V, T) \qquad \cdot \quad \cdot \quad \cdot \qquad 6$$

where G_o is constant (strain energy release rate in the limit of zero rolling speed) and $\psi(V,T)$ represents the overall dissipation function involving rolling speed, V, and temperature, T. The speed dependence is clear in Figs. 3 and 4. Let us now consider the temperature dependence.

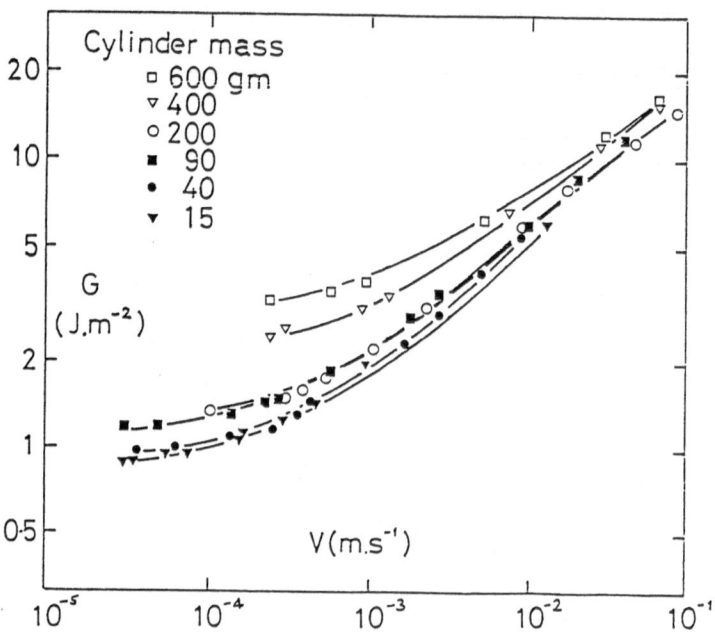

Figure 4   Results of G vs. V for NBR track and various cylinder masses.

Rolling tests were effected at various temperatures and examples of results are shown in Figs. 5(a) and 6(a) respectively for EPDM and NBR tracks. Cylinder weights were 170 gm and 300 gm. (Weights must be high or $\propto \phi$ will be completely masked by W). Clearly an increase in temperature leads to a decrease in G (at given V). Application of the WLF law allows us to express G as a function of reduced speed, $a_T V$, where a_T represents the shift factor calculated from ref. 10:

$$\log a_T = \frac{-C_1(T - T_g)}{C_2 + (T - T_g)} \qquad \cdots \quad 7$$

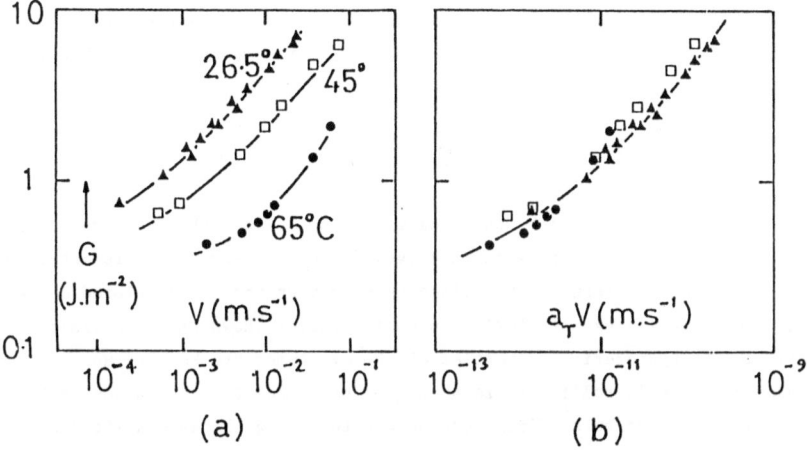

Figure 5   a) G vs. V at various temperatures for a
                cylinder of mass 170 gm. on EPDM track,

            b) Same results expressed as G vs. $a_T V$.

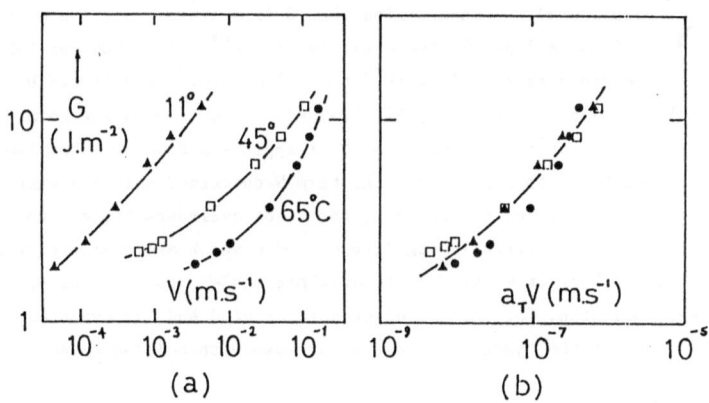

Figure 6   a) G vs. V at various temperatures for a
                cylinder of mass 300 gm. on NBR track.

            b) Same results expressed as G vs. $a_T V$.

The constants C_1 and C_2 may be taken as 13.1 and 40.7 for EPDM, and 8 and 39.5 for NBR [11]. The term $a_T V$ represents an equivalent speed evaluated at the glass transition temperature, T_g, of the elastomer (see Table 1). We can now write:

$$G = G_0 \psi \left(a_T V \right) \qquad \cdot \quad \cdot \quad \cdot \qquad 8$$

Figures 5(b) and 6(b) show log G vs. log $a_T V$ for the two elastomers in question and it can be seen that satisfactory superposition is obtained. The fact that the results at various temperatures can be correlated and are then found to lie satisfactorily on master curves shows that the visco-elastic, or hysteretic, properties of the elastomer tracks play an important rôle in rolling resistance. We have made the assumption that both adhesion and bulk effects can be treated using the same shift factor for a given polymer. Although this may be reasonable, it has not been verified and any differences could at least partially explain the scatter (albeit fairly small) observed in Figs. 5(b) and 6(b).

5. INTERPRETATION AND DISCUSSION

Considering equations 3 and 5 and assuming that both E and α remain relatively constant for the range of rolling speeds studied (and at a constant temperature), we should find that G is a linear function of (mg cos θ) $^{\frac{1}{2}}$. Figures 7 and 8 give G vs. (mg cos θ) $^{\frac{1}{2}}$ for EPDM and NBR at 20°C over the speed range 3.16×10^{-5} to 10^{-1} m. sec.$^{-1}$ and it can be seen that a reasonable linearity is obtained. By extrapolating to m --- 0, the bulk dissipation term of equation 5 disappears and we can thus isolate, at least to a first approximation, the term W concerned uniquely with adhesion hysteresis. As can be seen, for light cylinders (15 gm) the bulk dissipation term is virtually negligible, and G and W are nearly equivalent. Figure 9 gives W vs. V on logarithmic scales, as obtained by the above extrapolation, for the two elastomers studied and, as expected, it can be seen that the hysteretic losses increase with rolling speed.

Having determined W, it remains to investigate the term $\alpha \phi$ related directly to bulk elastomer dissipation. In order to do this, a technique suggested by Roberts was employed [12]. By covering the elastomer track with a thin layer of talc, the tendency for the steel cylinder to adhere

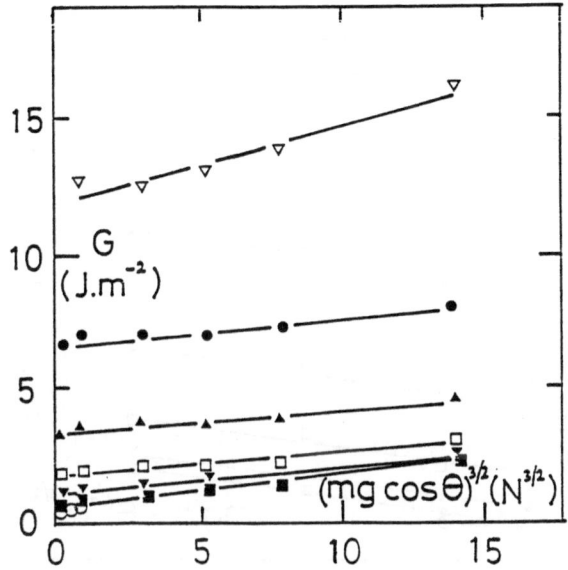

<u>Figure 7</u>   G vs. (mg cos θ)$^{3/2}$ for EPDM at rolling
speeds from 3.16 x 10^{-5} to 10^{-1} m. sec^{-1}.

becomes virtually zero and the value of obtained experimentally is almost
entirely due to the bulk dissipation term. Nevertheless, the interpretation of results on a non-adhering track is a little more delicate since the
cylinder no longer rolls at constant speed but accelerates.   Let us
consider the case where no dissipation whatsoever occurs. If the cylinder
starts from rest, after rolling unit distance down the track, the loss of
potential energy, mg sin θ will be equivalent to the gain in kinetic energy

$$mg \sin \Theta = mv^2/2 + I\omega^2/2 \quad \cdot \quad \cdot \quad \cdot \quad 9$$

where v and ω represent linear and angular velocity after unit rolling
distance and I is the cylinder moment of inertia. We can thus calculate
the expected terminal speed of a rolling cylinder over the length of track
to be considered. In addition, the actual terminal speed may be assessed
from measurement of rolling time and application of Newton's second law
of motion. This in turn allows us to assess the final kinetic energy.
The difference between the calculated kinetic energy without dissipation

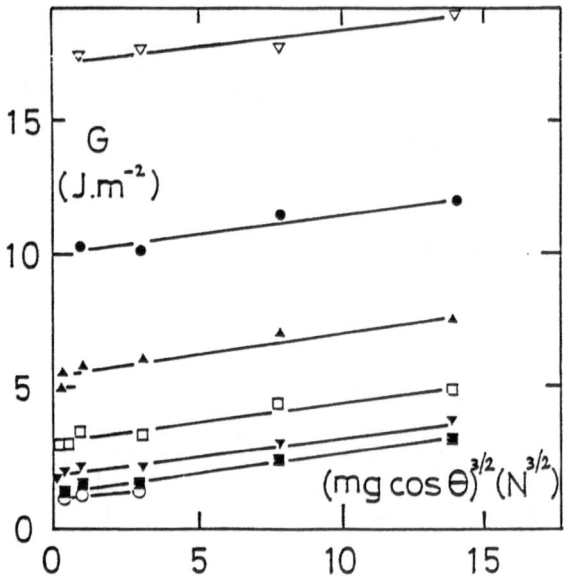

Figure 8   G vs (mg cos 0)   for NBR at rolling speeds
            from 3.16 x 10^{-5} to 10^{-1} m. sec^{-1}.

and that obtained experimentally amounts then to the dissipation term over
the entire track length, which, given the range of speeds considered, will
be almost entirely attributable to hysteresis in the elastomer as air
resistance will be negligible.   We can now assess $\propto \phi$.   There remains
nevertheless a problem.   Previously we have considered G vs. V where V is a
constant rolling speed.   However, when the cylinder accelerates, all speeds
between zero and the terminal speed are covered.   We thus have to consider
an average value of $\propto \phi$ over a speed range.   Should we take the average
speed or the terminal speed?   In all probability we should take something
between the two (since dissipative effects increase quite drastically with
speed, it is clear that we do not require a value less than the average
speed).   Fortunately we are interested in results on a logarithmic scale.
Since the final speed is simply twice the average speed (assuming constant
acceleration), the uncertainty produced between the two is simply log 2 or
ca. 0.3.   We have in fact adopted the average speed.

Having now estimated separately both the W and the $\alpha\phi$ contributions to
G of equation 5, we may attempt to reconstruct the curves for G and compare
them with the experimentally obtained results for heavy cylinders on elastomeric tracks without a talc layer. Figure 10 gives such a comparison for
the case of a 400 gm cylinder on both EPDM and NBR tracks.   Open symbols
correspond to experimental rolling data without talc.   Filled symbols
represent the addition of W (from Fig. 9) and $\alpha\phi$ from experiments on talc-
covered tracks. The agreement between the two approaches is very satisfactory and confirms that the proposed equation 5 is valid for explaining the
behaviour of a heavy cylinder rolling down an elastomeric track where
dissipative losses are related both to hysteretic effects at the elastomer-
cylinder separation front and to bulk viscoelastic phenomena occurring
within the elastomer and due to the internal stress field.

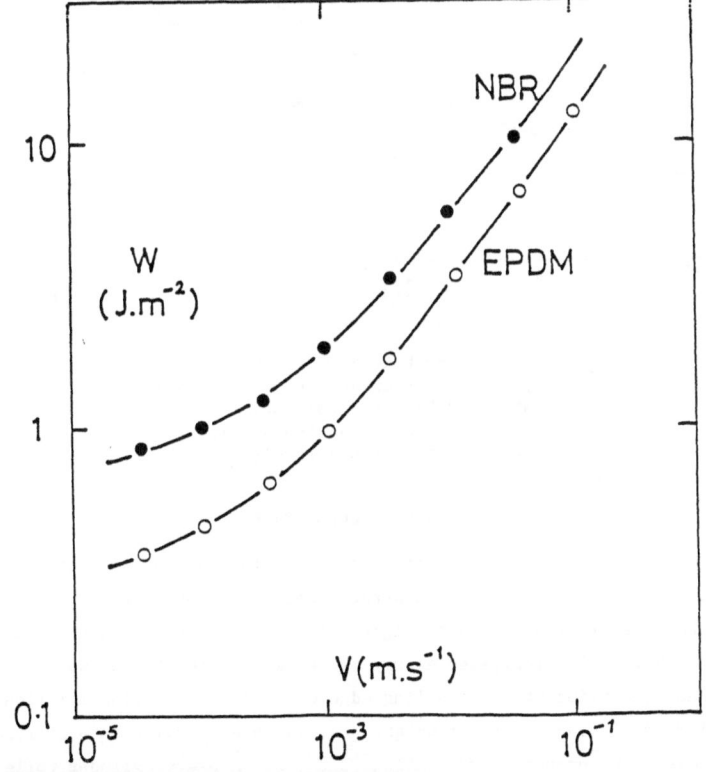

Figure 9   W vs V for both EPDM and NBR tracks, as
            obtained from Figs. 7 and 8.

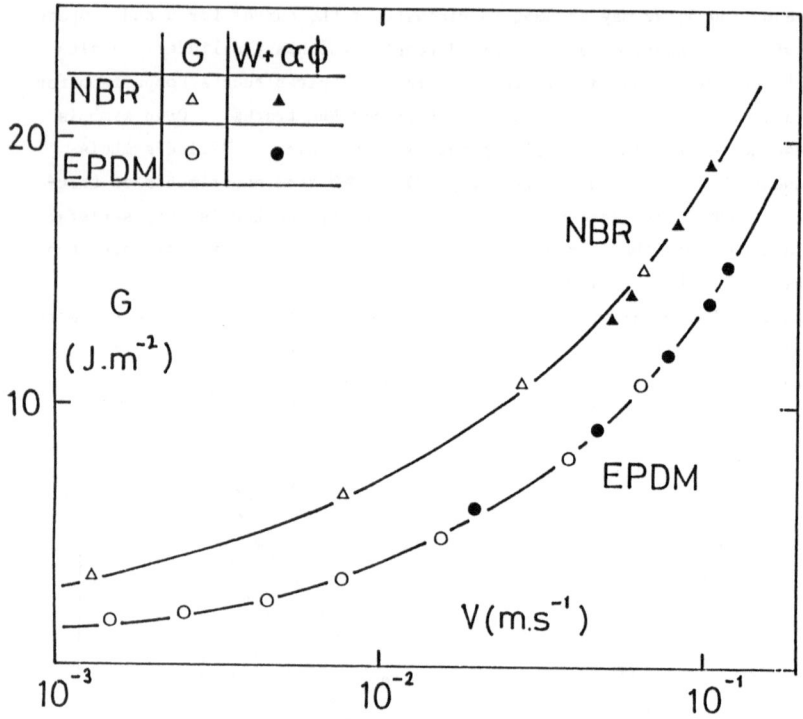

<u>Figure 10</u> G vs. W for both EPDM and NBR tracks and cylinder mass of 400 gm. Open symbols – data from "normal" rolling, filled symbols – addition of calculated W (Fig. 9) $+\alpha\phi$ (talc-covered track) results.

6. CONCLUSIONS

Rolling cylinder tests have been used in the past as suitable methods of studying elastomer-rigid substrate adhesion. Nevertheless, most work has been restricted to use of light cylinders for which the totality of viscoelastically dissipated energy can be attributed to the separation process occurring at the trailing edge of contact. Rolling friction of cylinders of non-negligible weight has also been studied in a different context. The present work combines the two. A heavy cylinder rolls down an elastomeric track and viscoelastic hysteretic losses are attributable both to the separation process and to the stress field beneath the cylinder and within the bulk elastomer. The two types of loss can be treated additively. Use of heavy cylinders can then be made for studying the adhesion of hard elastomer for which sufficient contact pressure is required to ensure good intimate cylinder-track contact.

The variation of rolling resistance with temperature has been considered and it has been shown that, at least to a first approximation, the separation and bulk hysteretic losses may be treated together. The WLF principle can be applied with success to obtain master curves of overall rolling resistance relating the strain energy release rate to corrected reference rolling speeds calculated at the glass transition temperature of the elastomer in question.

ACKNOWLEDGEMENTS

Thanks are due to the French Ministry of Research and Technology for financial support and to CETIM (Nantes) CPIO and Manurhin for their assistance.

REFERENCES

1. Roberts, A.D., Thomas, A.G., Wear, 33, 45, 1975.

2. Dupré, A., 'Theorie mécanique de la chaleur', Gauthier-Villars, Paris, 369, 1869.

3. Gent, A.N., Schultz, J., J. Adhesion, 3, 281, 1972.

4. Carré, A., Schultz, J., J. Adhesion, 17, 135, 1984.

5. Lake, G.J., Thomas, A.G., Proc. Roy. Soc. (London), A300, 103, 1967.

6. Greenwood, J.A., Minshall, H., Tabor, D., Proc. Roy. Soc. (London), A259, 480, 1961.

7. Hertz, H., J. Rein. Angew. Math., 92, 156, 1881, reported in Timoshenko, S.P., Goodier, J.N., 'Theory of Elasticity', 3rd edition, McGraw-Hill, New York, 414, 1970.

8. Johnson, K.L., Kendall, K., Roberts, A.D., Proc. Roy. Soc. (London), A324, 301, 1971.

9. Mangis, D., Barquins, M., J. Phys. D. Appl. Phys., 11, 1989, 1978.

10. Williams, H.L., Landel, R.F., Ferry, J.D., <u>J. Amer. Chem. Soc.</u>, **77**, 3701, 1955.

11. Ferry, J.D., <u>'Viscoelastic Properties of Polymers'</u>, Wiley, New York, 1970.

12. Roberts, A.D., <u>Rubber Chem. Technol.</u>, 54, 944, 1981.

A MODIFIED THERMOPLASTIC ADHESIVE
LAYER IN LAMINATES

R. J. Ashley
currently employed in the Product Innovation Section
at Metal Box plc, R&D Division,
Wantage, Oxon. OX12 9BP,
UK

1. Introduction.

The process of lamination is valuable to many
industries since it provides the opportunity to combine the
properties of differing materials to meet specific end uses.
Particularly useful is the ability to combine plastic and
metal substrates for such applications as building panels
with decorative finishes, weight reducing panels for the
aircraft and automotive industries, cable wrapping and
packaging.

Within the packaging industry, for example, thermoplastic films are used extensively. However, for some
applications, individual thermoplastic films are deficient
in certain properties such as permeability, processability,
heat sealability or rigidity. It is therefore useful to
laminate these films to other materials to provide a
structure that may be converted into a total package.
Combinations may be of different films each providing a
specific property or with the inclusion of metal foil or
paper. To be useful as a packaging material the conversion
process often necessitates a heat sealable layer. To achieve
this a relatively low melting polymer film such as a
homopolymer of ethylene or copolymer with vinyl acetate
would be laminated to a film providing good abuse resistance
and printability such as polyester, nylon and oriented
polypropylene, or to a barrier layer such as aluminium foil
plus supporting films.

The traditional techniques for the manufacture of
laminates are the use of cross-linkable adhesives, extrusion
coating, coextrusion or thermal bonding. In recent times
there has been concern over the use of solvents in adhesives
such that controls for retained solvent in the laminate and
emission to the atmosphere are necessary. The retained
solvent may lead to adhesion problems during the life of the
laminate such that off-odours or taint may be transmitted to

the product, or migrating polymer additives carried through
the structure. Bonding of the multi-layer laminate is
generally achieved by using a thermosetting adhesive of the
urethane type cross-linked by an isocyanate. Such an
adhesive is generally satisfactory for attaining a high bond
strength between various plastic and metal substrates, but
may contain some ingredients which can be extracted by
foodstuffs and be considered undesirable, especially where
the laminate is subjected to high temperatures. The
thermosetting resin may comprise uncondensed monomers and
low molecular weight polycondensates which tend to migrate
into or react with the packed product. In addition, whilst
high bond strengths may be initially achieved, flexing of
the laminate can often cause localised de-bonding which
could initiate more severe delamination of the plies and
lead to loss of container integrity. The cross-linking
reaction is time dependent and periods of 1 to 5 days,
possibly at elevated temperatures, may be required to attain
full bond strength. The extrusion lamination route gives
a more instantaneous bond, but is specific to where a
polyethylene or derivative can be used as the adhesive
layer.

In order to reduce the environmental pollution
problem of solvents there have been developments in the basic
urethane adhesive chemistry during the late 1970's and
early 1980's [1] .

Whilst laminates of dissimilar materials may have
desirable characteristics, it can prove difficult to
achieve satisfactory bonds between some substrates. It is
often difficult to bond polyethylenes to other substrates
due to differences in physical and chemical structures.
Bond improvements can be made by the use of surface
treatments such as corona discharge or primers, but are
subject to weakening by exposure to moisture or aggressive
products. An understanding of the improvements brought
about by surface treatments is emerging due to examination
by modern analytical techniques such as ESCA. Adequate
preparation and laydown of the adhesive and proper
treatment of the substrates is essential to avoid adhesion
problems [2] .

To overcome some of these difficulties the polar
content of a polyolefin may be directly increased by
grafting functional groups onto the polymer backbone chain.
The most common combination is maleic anhydride grafted
onto polypropylene although other variations are of interest.
Details of some of these materials, their manufacture and
how they may be used form the rest of this paper.

2. Nature of grafted polymers.

Some important factors to consider when producing a polyolefin-based adhesive are:-

(a) nature of polar groups to be introduced into polyolefin which improve their adhesive property,
(b) method of introducing polar group into polyolefin,
(c) molecular structure of polyolefins subject to modification,
(d) low molecular weight materials included in or related to polyolefins which inhibit adhesion,
(e) compounding technology.

Among these factors, the nature of polar groups and their method of incorporation have been found most important.

In the modified olefin resin under consideration ethylene monomer units may be included in the form of a random or block copolymer with propylene or in the form of a polymer blend or a combination.

A modified olefin resin may be prepared by incorporating a known carbonyl group containing ethylenically unsaturated monomer into a main chain or side chain of an olefin resin by conventional means such as graft or block or random copolymerisation or terminal treatment.

As the carbonyl group-containing ethylenically unsaturated monomer there can be used monomers having a carbonyl group ($\geq$C$=$0), derived from a carboxylic acid, an acid salt, anhydride, ester, amide or imide, or an aldehyde or ketone singly or in combination with monomers having a cyano group ($-$CN), hydroxy group, ether group or oxirane ring ($-$C$\underset{\diagdown O \diagup}{\underline{\quad}}C-$).

The carbonyl group-containing monomers may be used singly or as a mixture of two or more forms. Suitable examples of monomers that may be used are as follows:-

(i) ethylenically unsaturated carboxylic acid monomers such as mono or polybasic carboxylic acid - acrylic, methacrylic, crotonic, fumaric, maleic, itaconic, citraconic acids;

(ii) unsaturated carboxylic anhydrides such as maleic, citraconic, tetrahydrophthalic anhydrides;

(iii) unsaturated esters such as ethyl acrylate, methyl methacrylate, 2-ethyl hexyl acrylate, monoethyl maleate, diethyl maleate, vinyl acetate, vinyl propionate, glycidyl methacrylate;

(iv) unsaturated amides and imides such as acrylamide, methacrylamide and maleimide;

(v) unsaturated aldehydes and ketones such as acrolein, methacrolein, vinyl methyl ketone, vinyl butyl ketone.

Typical thermoplastic resins suitable for modification by grafting with the above include the following:-

(i) high density polyethylene,
(ii) polypropylene,
(iii) polyamides,
(iv) polyethylene terephthalate,
(v) ethylene/propylene copolymers.

In addition, blends of these resins with lower softening point resins such as low density polyethylene, ethylene/acrylate copolymers, ethylene/vinyl acetate copolymers, random ethylene/acrylic or methacrylic acid copolymers, ionomers, chlorinated polyethylene can be used.

The carbonyl group-containing monomer is bonded to the main or side chain of the olefin resin so that the carbonyl group concentration is within the range 0.5 to 15 weight percent, and the degree of crystallisation of the resulting modified olefin resin is at least 15%.

Researchers have concluded that maleic anhydride is the most significant modifying agent and grafting onto the backbone of a polymer chain the best method of incorporation. The most common backbone polymers used are propylene homopolymer and ethylene-propylene random or block copolymers.

A typical structure for a maleic anhydride grafted polyolefin or maleated polyolefin is shown in Fig.1.

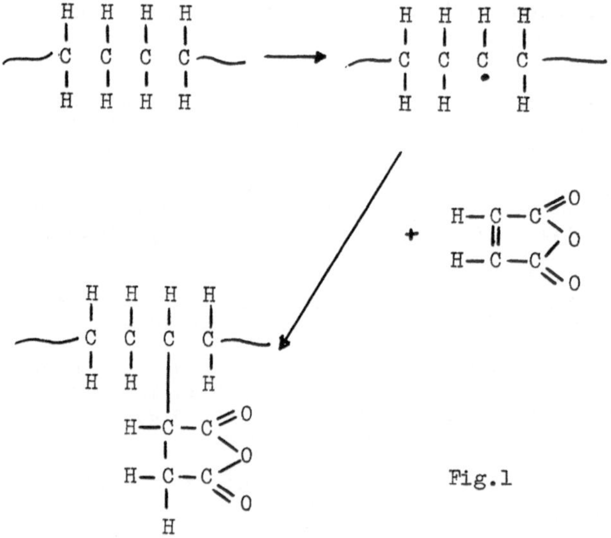

Fig.1

In the specific case of a modified propylene polymer produced commercially the solid resin will contain about 0.5 to 4% chemically combined maleic anhydride by weight of the polymer.

Another commercial material is a graft of acrylic acid onto ethylene polymer. The amount of graft unit varies between 3 and 15% by weight. The low acid resins are useful for extrusion coating whilst the high acid resin is dispersible in alkaline water to provide an aqueous adhesive or coating for metal substrates.

Standard ethylene homopolymers exhibit good flexibility, moisture barrier and tensile properties. They are crystalline in nature and completely non-polar, hence have poor optical and adhesion properties. However, copolymerisation with acrylic acid significantly alters the molecular structure by random inclusion of bulky ionic carbonyl groups along the backbone and side chains ( ).

$$\underset{H}{\overset{C}{\underset{O}{\diagup}}\overset{}{\diagdown}O}$$

The presence of the carbonyl group brings about important changes;

(i) the carboxyl groups are free to form bonds and to interact with any polar substrate - metals, cellulosics, polyamides; - the increase in adhesion is roughly proportional to the acrylic acid content.

(ii) carboxyl groups on adjacent chains can hydrogen-bond with each other providing an internal toughness to the polymer.

(iii) the bulky side groups inhibit the ability of the polymer to crystallise improving clarity and reducing softening point. The materials thus exhibit good hot tack at low melting temperatures.

At acid levels in excess of 15% the copolymer may be dispersed in alkaline water to provide a very small particle size colloidal dispersion. Conventional coating and laminating equipment can be used to apply the aqueous dispersion as a primer or adhesive. Further modification of the coating is possible by blending with reactive cross-linking agents such as urea-formaldehyde resins.

3. Manufacture of graft polyolefin.

The modified polymer can be prepared by reacting maleic anhydride or the like with any solid propylene polymer containing active sites which are capable of anchoring the graft onto.

Active centres at which anchoring can occur are induced into the polymer chain by the following methods:-

(a) by subjecting the polymer to the action of high energy ionising radiation such as gamma rays, X-rays, or high speed electrons;

(b) contacting the polymer either as a solid or as a solution in solvent with a free radical producing material such as dibenzoyl peroxide;

(c) milling the polymer in the presence of air or blending in the melt phase in an extruder or ultrasonic vibration.

In the case of reaction in a solution system, the olefin resin, monomer and initiator are dissolved in an aromatic solvent such as toluene or xylene and the grafting reaction carried out. The resulting modified olefin is recovered as a precipitate and purified by washing or extraction so as to remove unreacted monomer, homopolymer or residual initiator. When the resulting modified olefin resin is recrystallised from an aromatic solvent and the crystallisation conditions appropriately controlled the particle size can be adjusted.

The method of incorporation of the modifying agent into the polymer can be varied according to the ultimate usage for which the resin is intended, determined by availability of equipment and nature of polymers, eg. pellets, fibres, solution or pre-formed film.

4. Adhesion mechanism.

There is no satisfactory explanation to fully account for the adhesion mechanism between a maleated polyolefin and other polymers, but the following points are relevant.

The nature of bonding forces involved between maleated polyolefins and other substrates is not clear. In some cases covalent bonds may be formed whilst in others, either ionic, hydrogen or Van der Waals bonds. In the case of bonding to polyamides or ethylene vinyl alcohol, commonly used in coextruded structures, a covalent bond is most likely. Study of the reaction between maleated polypropylene and ethylene vinyl alcohol in the molten state reveals ester bond formation by IR spectroscopy, with greater ester bond formation for higher concentrations of the acid anhydride group. The reaction for the bond formation may be represented by Fig.2.

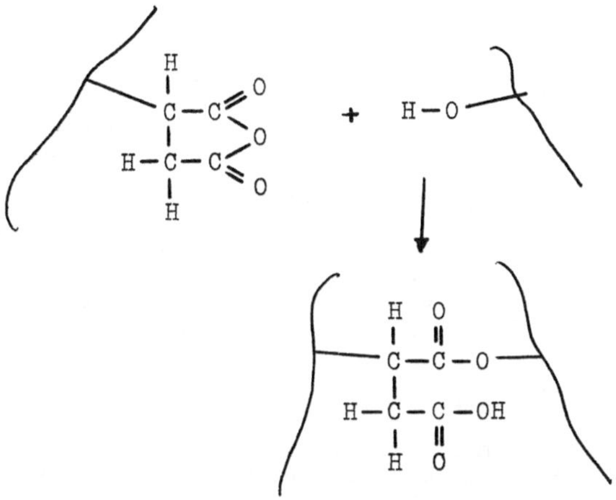

Fig.2

Interfacial amounts of polar groups can be examined by ESCA. In an example of maleated polypropylene contacted in a molten state with aluminium foil, the interfacial concentration of oxygen belonging to the acid anhydride group is higher than in the internal region of the adhesive layer. This supposes that during processing where maleated polyolefin contacts with a polar material such as polyamide, ethylene vinyl alcohol or metal the polar group diffuses towards the adhesive interface. This may explain how maleated polyolefins, with a relatively small amount of polar group, can exert strong adhesive bonds.

Adhesion strength is decreased by the formation of a weak boundary layer and internal stress at the interface. Similar observations can be found with the modified thermoplastic adhesive layers. When slip additives or surface modifiers are needed in the adhesive formulation adhesive strength is decreased significantly. This is due to the additives diffusing more easily to the interface than the grafted acid anhydride group and forming a weak boundary layer in the interface. Internal stress becomes a significant problem when the adhesive and substrate are relatively stiff, however, the ability to blend the modified adhesive polymer with more flexible materials can contribute to reduced internal stress and maintained adhesive strength.

Thus the adhesive strength between the modified polymer adhesive and polar substrates depends upon the degree of chemical interactions in the interface. Since development of chemical reactions and diffusion of polar groups to the interface are dependent on temperature and time the

processing conditions can have a critical impact.   Conditions
imposed may include the following:-

    (a) extruding temperatures of the adhesive polymer;
    (b) contact time and pressure in the molten state;
    (c) pre or post heating of the substrate;
    (d) deformation of the adhesive-substrate layers,
        reducing concentration of interfacial chemical bonds.

    In general, the higher the extrusion temperature (but
below a maximum likely to cause decomposition of the polymer),
and the longer the molten contact time, the higher the
adhesion strength that can be obtained.   To increase the
contact time the use of pre or post heating cycles to the
substrates or laminate may prove beneficial.

5. Methods of application.

    The modified polymers described may be used in a
variety of ways depending upon equipment availability.   A
few examples are given below.

    (i) Extrusion coating – A widely used conversion process
to produce a laminate of metal foil and polyethylene.
However, to produce adequate bonding it is often necessary
to prime the foil.   The modified polymers may be used from
pellet form and extruded directly onto the metal foil or
other heat resistant substrate without ancillary processes.
The resulting laminate generally exhibits improved product
resistance.

    (ii) Coextrusion coating – the modified layers may be used
as a tie layer between metal foil and a polymer film, eg.
metal/ethylene acrylic acid/LDPE.   Rather than extruding
the modified polymer between the other substrates it is
possible to coextrude a thin layer of the EAA resin onto
the base or carrier film.   The combined resins could be
coextrusion coated or prepared as a multi-layer structure
with a polar surface for subsequent bonding, eg. by thermal
lamination in a high temperature and pressure nip.   To
improve the bonding a pre-heated metal substrate may be used.

    (iii) Powder coating – The powdered polymer may be coated
by electrostatic spraying or fluidised bed techniques.   This
is especially useful for application to irregular shaped
articles.   Instead of coating a solid powder and then melting
the powder to form a continuous film, the resin may be
coated in the form of a molten powder using a flame spray
method.   A film of polymer which is compatible with the
modified resin may then be laminated to the coated layer,
preferably by extrusion coating and bonded under heat and
pressure to obtain the desired laminate.

    (iv) Dispersion coating – The modified resin may be
dispersed in water or a low-boiling organic solvent and the

dispersion coated onto a metal substrate. After evaporation of the solvent a powder state results and the temperature must be raised to fuse the powder into a continuous film.

(v) Adhesive films - the modified resin may be prepared as a roll of film. This can subsequently be used as an insert between metal foil and other polymer films for thermal lamination by compression in a hot roller system. Either a continuous lamination process or batch process combining sheets in a hot press can be used to assemble laminates.

(vi) Hot melt adhesives - the modified resins may be used in hot melt adhesive formulations to bond metals, paper and paperboard, cellophane, leather, nylon and polyethylene. The resins exhibit good hot tack properties at low melting temperatures.

6. Summary.

This paper has briefly indicated how a relatively inert polymer such as a polyolefin may have its adhesive properties enhanced by grafting certain polar groups to the polymer backbone. By doing this it may be possible to avoid the use of other adhesives as a separate application process to assemble a laminate. In addition, lengthy cure periods associated with some adhesives may be avoided. Also demonstrated are the wide variety of conversion processes that may be used with the modified resins to produce commercially viable laminates.

REFERENCES.

1. Ashley,R.J., A review of adhesives for the flexible packaging industry during the 1980's. In Adhesion 7 , ed. K.W.Allen, Elsevier Applied Science Publishers, London, 1983, pp. 221-251.

2. Ashley,R.J. et al., Adhesion problems in the packaging industry. In Industrial Adhesion Problems, ed. D.Brewis and D.Briggs, Orbital Press, Oxford, 1985, pp. 199-257.

THE USE OF ELECTRON MICROSCOPY FOR THE ANALYSIS OF THE
ADHESIVE-ADHEREND INTERFACE IN THE ALUMINIUM-ALUMINIUM BONDED JOINT

J.A.Bishopp*, E.K.Sim, T.V.Smith, G.E.Thompson and G.C.Wood

*Bonded Structures, Ciba-Geigy Plastics, Duxford, Cambridge CB2 4QD
Corrosion and Protection Centre, University of Manchester
Institute of Science and Technology, Manchester M60 1QD, England

INTRODUCTION

Over the past 25-30 years much theoretical work has been carried
out in an attempt to analyse the adhesive bond and hence determine the
nature of the mechanisms of adhesion.

Current theories include those of Zisman and co-workers[1] where
intimate wetting of the adherend by the adhesive is required not only
to facilitate the formation of covalent bonds between adherend and
adhesive, but also to allow the inter-molecular attraction forces (Van
de Waal and, to a lesser extent, the London dispersion forces) to come
into play.

Other theories range from the disappearance of a definite interface between adhesive and substrate with the formation of interpenetrating networks [2,3] (diffusion theory) to the electrostatic
theories of Derjaguin et al[4]. For the latter, electron transfer on
contact between adhesive and adherend, result in the formation of a
double layer of electrical charge; adhesion is due to the attractive
forces across this electrical double layer.

(Many researchers believe that this physico-chemical adhesion can
be enhanced by simple mechanical keying and have thus, advanced the
theory of "mechanical" adhesion; the adhesive supposedly fills and
forms round surface discontinuities in the substrate and then, on
hardening, forms a strong mechanical key, locking the two substrates
together).

The various theories have essentially been based on the examination of model systems coupled with an in-depth study of the physico-
chemical properties and thermodynamics of the adhesive, the adherend
and the adhesive-adherend interface. Such analyses, of necessity,
only consider a distance of about 1-2nm across that interface. Thus,
whilst giving an insight into the mechanisms of adhesion, they do not
give the adhesive formulator much information about the way in which
the adhesive will perform within an actual, load-bearing, bonded
structure. Nor, at the other end of the scale, does a simple visual
inspection of a bonded and/or ruptured joint yield much useful data.

Hence, the ability to use a "visual" method to analyse the joint,
particularly after rupture, close to the original interface between
adhesive and substrate, would represent a significant advance.

Nowadays, the use of scanning electron microscopy (SEM) has become well established for examining both the surface of an adherend and the adhesive fracture surface within the ruptured joint, but this still gives little information about the actual interface.

With the advent of improved ultramicrotomy techniques (5), where 10-20nm thick slices can be taken through both substrate and adhesive and then viewed by transmission electron microscopy (TEM), the possibility of making a direct examination of the adhesive-adherend interface has become a reality.

Here, the initial findings are reported of a major programme which is making use of SEM and TEM techniques not only to characterise substrate surfaces actually used by the Adhesive Industry but, hopefully, to obtain a better understanding of the adhesive joint under various environmental conditions.

RESULTS AND INTERPRETATION

Surface Analysis of the Substrate

The range of substrates which the Adhesives Industry requires to be bonded is vast: wood, glass, plastics, metals, rubber, fibres, composite materials (e.g. CFRP), etc. One of the most common adherend materials currently encountered when bonding structural (i.e. load bearing) components is aluminium and it is this material - specifically Alclad 2024-T3 (Table 1), a grade much used in the Aerospace Industry and having something of a reputation of being difficult to bond - which has been used for the work reported here. Later, it is intended to extend the analysis to encompass unclad aluminium alloys of different compositions, including those based on aluminium-lithium.

When bonding structural components the adherends are usually given some form of pretreatment to ensure that the adhesive is applied to a clean, stable surface so that as good a bond as possible can be formed. However, economic and end-use considerations often mean that a less-than-optimum form of pretreatment, and occasionally no pre-treatment, is used.

The first stage of this work has been to characterise the surfaces formed following the more usually encountered pretreatments:-

i) Solvent wash + "light" abrade using Scotchbrite$^{(R)}$ 96 or wire wool + clean and rewash.

ii) Solvent wash + shot-blast using a 180-250 micron alumina grit + air-blast, brush clean and rewash.

iii) Trichloroethylene vapour degrease + chrome/sulphuric acid pickle to DTD 915B(ii)(6).

iv) Vapour degrease + DTD 915B(ii) pickle + chromic acid anodise to DEF 151(7).

v) Vapour degrease + DTD 915B(ii) pickle + phosphoric acid anodise to BAC 5555 (8,9).

Simple, high-resolution SEM examination of the "mechanically" pretreated surfaces reveals sufficient data to permit their characterisation.

Figure 1 shows the effect of light abrasion (Scotchbrite). A very rough "furrowing" through the aluminium cladding has taken place with some of the ploughed metal remaining loosely attached to the lip of the furrow. This represents potential areas for failure initiation in the bonded joint. A further cause for concern is also revealed (Figure 2) where relatively small pieces of abrading material can be seen embedded into the roughened surface. These inclusions can build stress concentrations into the adhesive joint which could cause premature failure. They can also act as possible sites for environmental attack.

The results of shot-blasting can be seen in Figure 3. The surface has been cut open giving a pseudo-plastic fracture profile. Examination at relatively high magnification (Figure 4) reveals considerable abrasion detritus remaining loosely bound to the surface; again such features are likely to cause premature failure.

Figures 5a - 7a show how little data, relative to their true morphology, surface examination of the chemically pretreated adherends does reveal. The expected etching craters are visible but other detail is probably masked by the sputtered gold coating.

TEM examination of replicated surfaces could have been employed (10, 11) but it is of limited use here as details of both the surface topography and morphology are required. Use of the ultramicrotomy/TEM technique is, however, ideal and yields the data sought.

Figure 5b shows that a surface film, with needle-like whiskers (about 30nm in length), is formed on chrome/sulphuric acid pickling of the aluminium adherend.

For chromic acid anodised adherends (Figure 6b), a "frond-like" columnar anodic film growth, to a thickness of about 2-5 microns, can clearly be seen; the surface topography mirrors the original, rough, pickled surface and there is little-to-no evidence of pore collapse.

Phosphoric acid anodising of the adherends (Figure 7b) leads to a similar "frond-like" anodic film. This time the thickness is about 0.8 micrometres. The surface topography is, however, considerably rougher than that for chromic acid anodising; a significant amount of pore collapse has led to a film surface showing the presence of cavities.

This technique, for surface characterisation, can also be used, for example, to examine the build up of the anodic film layer with time.

TABLE 1: Composition of ALCLAD 2024-T3

	Percentage Composition	
	Core	Cladding
Copper	3.8 - 4.9	0.1
Magnesium	1.2 - 1.8	-
Manganese	0.3 - 0.9	0.05
Iron	0 - 0.5	)
Silicon	0 - 0.5	) 0.7
Chromium	0 - 0.1	-
Zinc	0 - 0.25	0.1
Others (individual)	0 - 0.05	0.05
Others (total)	0 - 0.15	0.15
Aluminium	Rest	Rest

TABLE 2: Variation in floating roller peel strength with time, for adhesive joints bonded with a 120°C curing toughened epoxy adhesive exposed to 85% RH at 70°C

Pretreatment of Alclad 2024-T3 Adherends	Peel Strength in N/25.4mm after			
	0 days	10 days	20 days	30 days
No Pretreatment	113	36	14	<10
"Light" Abrasion	150	75	24	22
Shot Blast	65	21	14	<10
Chrome/Sulphuric Pickle	227	155	132	105
Chromic Acid Anodise	240	152	87	40
Phosphoric Acid Anodise	298	232	183	202

Surface Characterisation of Mechanically Abraded Alclad

 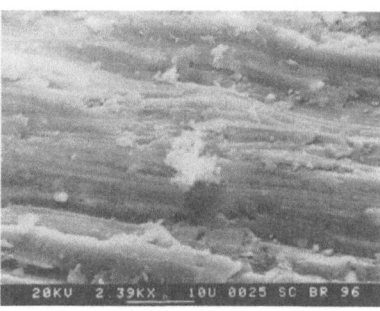

Figure 1 : (Bar = 50 microns)   Figure 2 : (Bar = 10 microns)

SEM Analysis of Scotchbrite Abraded Alclad

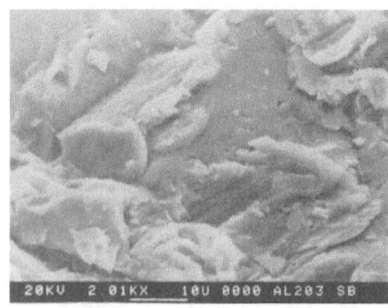 

Figure 3 : (Bar = 10 microns)   Figure 4 : (Bar = 1 micron)

SEM Analysis of Shot-blasted Alclad

253

Surface Characterisation of Chemically Pretreated Alclad

Chrome/Sulphuric Acid Pickle

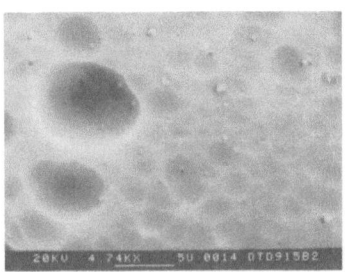

Figure 5a :SEM Analysis
(Bar = 5 microns)

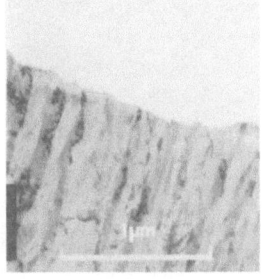

Figure 5b : Ultramicrotomy/TEM
Analysis (Bar = 1 micron)

Chromic Acid Anodise

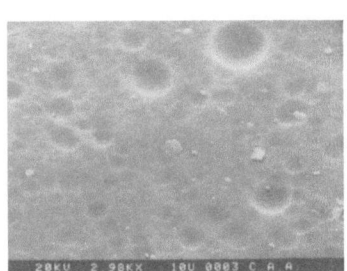

Figure 6a : SEM Analysis
(Bar = 10 microns)

Figure 6b : Ultramicrotomy/TEM
Analysis (Bar = 1 micron)

Phosphoric Acid Anodise

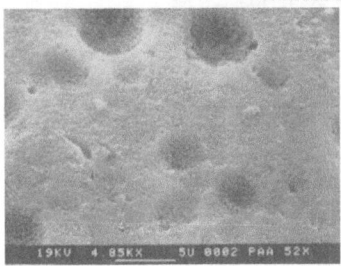

Figure 7a : SEM Analysis
(Bar = 5 microns)

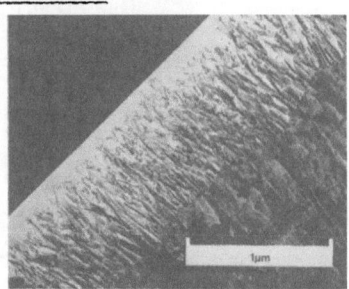

Figure 7b : Ultramicrotomy /TEM
Analysis (Bar = 1 micron)

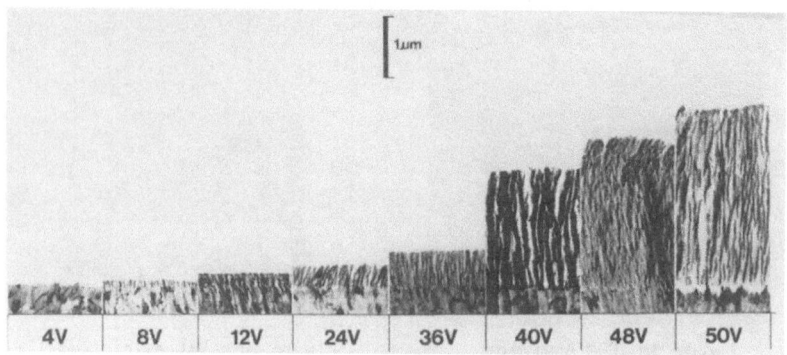

Figure 8 : Growth of Anodic Film with Time and Voltage

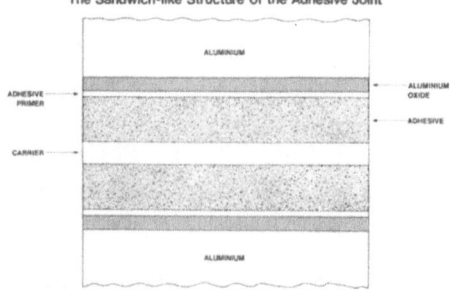

Figure 9 : Schematic Representation of the Adhesive Joint

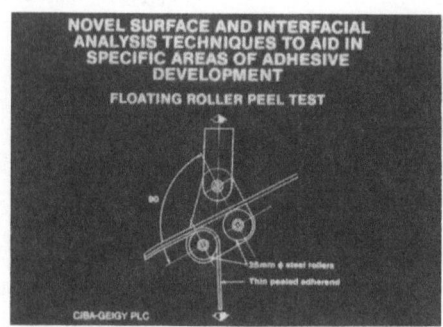

Figure 10 : Schematic Representation of the Floating-Roller Peel Test

In accordance with DEF 151, the anodising voltage is slowly stepped up from 0-40 Volts (4V/minute), a 20 minute hold at 40 Volts is followed by a 2V/minute step up to 50V and this voltage is held for a further 5 minutes. Figure 8 shows that, for Alclad 2024-T3, relatively little anodic film growth takes place until 40 Volts is reached, thereafter growth is rapid. The need, therefore, rigidly to control bath conditions and voltage increase rates is essential to the production of reproducible anodic oxide films.

The Adhesive Joint

Characterisation of the adherend surface is relatively simple to accomplish but when the adhesive joint is considered the whole picture becomes more complex. Figure 9 shows that the joint is essentially a sandwich structure.

For this work, the joint construction, usually 25mm-wide floating-roller peel specimens (Figure 10), was as follows:-

i)   Aluminium : Alclad 2024-T3 pretreated as described above.

ii) Primer : Generally the aluminium was unprimed. However, two specific areas did call for their use. Firstly, titanate-labelled epoxy primers have been used to aid in determining the depth of adhesive penetration into the anodic film. Secondly, standard epoxy-phenolic corrosion-inhibiting primers are currently being used in the studies on the environmental resistance of the joint.

iii) Adhesive : Toughened epoxy formulations in film form. Currently, adhesives curing at 120°C are being used but, at a later date, it is intended to evaluate 150 and/or 170°C curing systems.

iv) Carrier : The adhesive film always contained a knitted nylon carrier to ensure good glueline thickness control.

As metallic adherends were being used a definable interface between adhesive and adherend would exist (i.e. - no interpenetrating networks) and hence, for this analysis, the area within the joint of prime interest was the interface between aluminium oxide film and adhesive or primer.

Interfacial Analysis in the Adhesive Joint

To prepare specimens for interfacial analysis the bonded joints were submitted to increasing peeling loads until failure occurred; examination of the failure modes at or around the interface could then take place.

Here, however, the use of ultramicrotomy/TEM analysis on its own is limited due to the small area of interface selected for examination; the chances of selecting a non-typical region are, therefore, significant. If,though, both low and high resolution SEM analyses of the fracture surfaces are coupled with the ultramicrotomy/TEM techniques, typical fracture areas can be identified and then characterised.

Figures 11 - 13 show that for the mechanically abraded surfaces relatively poor adhesive bonding has taken place with a high degree of voiding or blistering being evident within the glueline.

When a film adhesive is laid up against an adherend for bonding, a layer of air is unavoidably left between the adhesive and adherend. The flow of the adhesive during the heat up to cure temperature has, therefore, a two-fold purpose; firstly to displace the trapped air and then to give an intimate wetting of the adherend to ensure good adhesion after cure. For a film adhesive there is, obviously, a finite time for air displacement and wetting, i.e. from melt to the point where cross-link insertion causes a significant increase in viscosity. In the case of lightly abraded and shot-blasted surfaces, it is likely that too rough a surface has been prepared, and not all the air can be displaced before onset of cure. From the character-isation of the surface pretreatment (Figures 2 and 3) it is clear that shot-blasting creates more new surface than abrading with Scotchbrite and, if the above is true, lower adhesive strength would be expected on these substrates. This is, in fact, borne out by the figures generated (Table 2).

The above is further confirmed when the ultramicrotomed sections are examined by TEM.

In Figure 14, significant areas of non-contact between adhesive and adherend (Scotchbrite abraded) can be seen.

Figure 15 shows that, whilst some of the major shot-blast craters are filled and wetted by the adhesive, the smaller ones are not. It also highlights the damage that can occur to the substrate. Figure 16 shows how loosely attached surface detritus can prevent wetting and also (centre picture) reveals what could be a stress crack due to the shot-blast operation itself.

The results of ultramicrotomy, therefore, indicate that it is not correct to say that, for all adhesives, the rougher the surface the better the adhesion. For the structural film adhesives, at least, it appears that it is intimate wetting rather than any mechanical keying which ensures good adhesion.

SEM analysis of the fracture surfaces generated using the chemical pretreated adherends show very different results.

Joints using DTD 915B (ii) pickled substrates (Figure 17) show "deep" cohesive failure throughout the adhesive layer. On chromic acid anodised substrates, however, significant cohesive failure of the adhesive only takes place in the bulk surrounding the nylon carrier; apparent adhesion failure to the adherend takes place in the carrier interstices (Figure 18); the expected cracking of the anodic film which occurs during peel testing, is clearly visible. On the other hand when phosphoric acid anodised substrates are used, a failure pattern is obtained which is much closer to that seen on pickled adherends (Figure 19).

SEM Analysis of the Fracture Surface of a
Ruptured Peel Joint
(Mechanically Abraded)

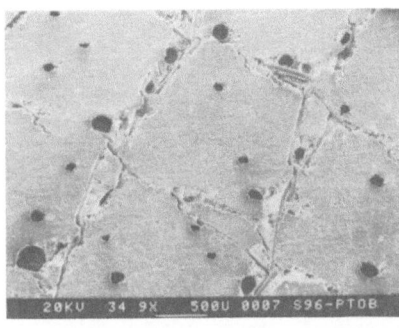

Figure 11 : Scotchbrite Abraded
(Bar = 500 microns)

Figure 12 : Scotchbrite Abraded
(Bar = 100 microns)

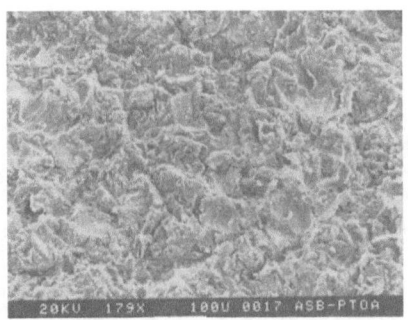

Figure 13 : Shot-blasted
(Bar = 100 microns)

--ooOoo--

Ultramicrotomy/TEM Analysis of the Ruptured Peel Joint
Mechanically Abraded)

Figure 14 : Scotchbrite Abraded
(Bar = 1 micron)

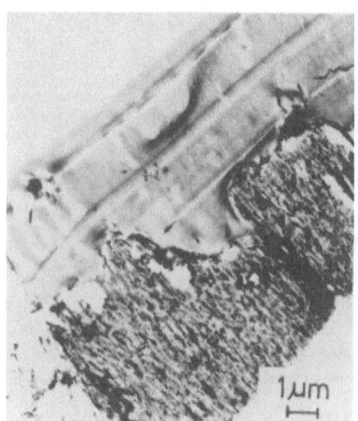

Figure 15 : Shot-blasted
(Bar = 1 micron)
Sub-surface damage can be
seen

Figure 16 : Shot-blasted
(Bar = 1 micron)
Looseley bound detritus
and possibly stress cracking
(centre) can be seen

--ooOoo--

259

SEM Analysis of the Fracture Surface of a Ruptured Peel Joint
(Chemically Pretreated)

Figure 17 : Chrome/Sulphuric Acid Pickle
(Bar = 200 microns)

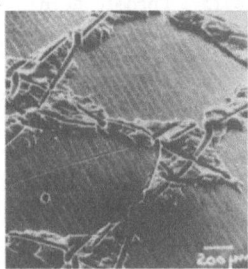

Figure 18 : Chromic Acid Anodise
(Bar = 200 microns)

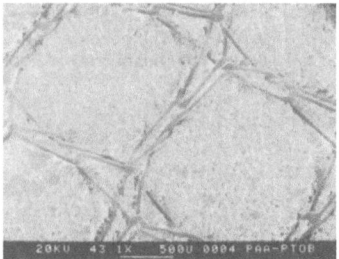

Figure 19 : Phosphoric Acid
Anodise (Bar = 500 microns)

--ooOoo--

Ultramicrotomy/TEM Analysis of the Ruptured Peel Joint
(Chemically Pretreated)

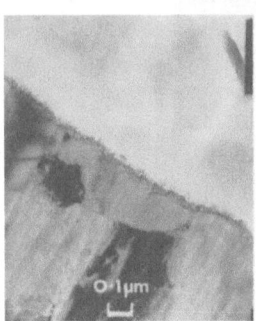

Figure 20: Chrome/Sulphuric
Acid Pickle (Bar = 0.1 microns)

Figure 21 :Chrome/Sulphuric
Acid Pickle (Bar = 1 micron)

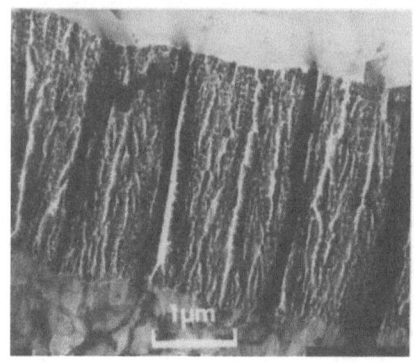 

Figure 22 : Chromic Acid Anodise
(Bar = 1 micron). Taken through
an area of "deep" cohesive
failure in the adhesive

Figure 23 : Chromic Acid
Anodise (Bar = 1 micron).
Taken through an area of
apparent adhesion failure

Figure 24 : Phosphoric Acid Anodise (Bar = 1 micron)

--ooOoo--

Specific Adhesion Problems

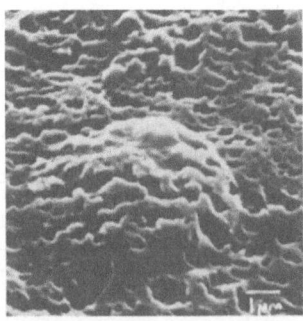

Figure 25 : SEM Analysis of the Rupture Surface of a Peel
Joint (Chromic Acid Anodised Adherends) -showing poor wetting
of adherend by adhesive

Figure 26 : Ultramicrotomy/TEM Analysis of Peeling Face from the
same Joint - showing true adhesion failure of adhesive to adherend.

Ultramicrotomy/TEM techniques allow a much more detailed characterisation to be made for these systems.

Figure 20 shows the excellent wetting of the film developed on pickled adherends. The effect of this good wetting can be seen in Figure 21 where some failure close to the oxide-aluminium interface has occurred.

A section through the area of adhesive cohesive failure for the chromic acid anodised specimen again shows excellent wetting of the anodic film; although it is impossible to tell whether any penetration down the pores has taken place (Figure 22). A section taken through an area of apparent adhesion failure (Figure 23) clearly shows adhesive still attached - thickness of about 100nm. Thus, although the failure, to the eye and to the scanning electron microscope, is a mixture of cohesive and adhesion failure it is, in reality, a true cohesive failure across the joint.

Sections through a phosphoric acid anodised specimen again show the extremely good wetting of the surface by the adhesive (Figure 24). Whilst it is obvious that the cavities caused by the pore collapse are filled by the adhesive once again it is impossible to tell whether any profound penetration down the pores has taken place.

DIAGNOSIS OF ADHESION PROBLEMS

The ability to distinguish between apparent and true adhesion failure (where there is genuinely no adhesive retained on the substrate) has already been the key factor in solving a specific adhesion problem (12, 13). The peel properties of an experimental structural adhesive were excellent when bonding pickled substrates but not acceptable when using chromic acid anodised adherends. High-resolution SEM analysis indicated that whilst the adhesive had flowed well enough to conform to the major surface topography, within these conformations, i.e. on a truly microscopic scale, it had not wetted sufficiently well to displace the entrapped air (Figure 25). These observations were confirmed by TEM studies of ultramicrotomed sections taken through the peeling face (Figure 26) where genuine adhesion failure to the anodic oxide film could be seen.

ENVIRONMENTAL STUDIES

This work is currently in its infancy but it is hoped to obtain a better understanding of environmental attack (especially high humidity and possibly salt-spray) in the bonded structure, by examining not only ruptured and non-ruptured areas immediately either side of the crack tip but also the crack tip itself for evidence of interfacial debonding, oxide hydration etc. A useful adjunct to this basic work will be any insight that might be gained into the mechanisms by which primers can protect the adhesive joint in these environments.

On unprimed surfaces, simple adhesion tests rate the pretreatments under consideration in the following descending order. Phosphoric acid anodised, pickled, chromic acid anodised, lightly abraded, shot-blasted (Table 2).

Both Auger and energy dispersive X-ray analyses have been
carried out on sections taken through the phosphoric acid and
chromic acid anodised adherends in an attempt to explain the very
different results. Whilst bound phosphates, which would give a
degree of protection, were present throughout the film formed by
phosphoric acid anodising, no similarly bound chromates were found
in the chromic acid anodised aluminium. The more open film developed
by anodising in chromic acid rather than pickling in chrome/sulphuric
acid could lead to easier environmental attack and, hence, could
explain the poorer results on the chromic acid anodised substrates
than on the pickled.

Work has already commenced on the interfacial analysis of these
systems and on joints prepared using corrosion inhibiting primers.
This will be reported at a later date.

CONCLUSIONS

Although in its early stages, the technique of coupling SEM
analysis of the surface with TEM examination of an ultramicrotomed
section taken through the same area is already generating much data
about both the pretreated adherend and the adherend-adhesive interface.
That a better understanding of the effects of pretreatment, the
environment, substrate and adhesive composition on the adhesive
joint will result is now assured.

ACKNOWLEDGEMENTS

The authors gratefully acknowledge the help of the following:

British Aerospace, Weybridge, for preparing the phosphoric acid
anodised adherends.

P.Silvester and G.Grigg, Ciba-Geigy, Central Research, Manchester,
for providing additional SEM data.

Dr.P.Wilford, RAE, Farnborough, for releasing details on RAE/UMIST
studies on pretreated aluminium adherends.

Mrs.S.C.Harrow and Miss J.A.Underwood for carrying out all the
mechanical testing and providing the specimens for SEM and TEM
work.

REFERENCES

1. Zisman, W.A., Industrial & Engineering Chemistry, 55, 1963, p 19.
2. Voyutskii,S.S., "Autohesion and Adhesion of High Polymers", John Wiley and Sons (Interscience), New York 1963.
3. Vasenin, R.M., "Adhesion, Fundamentals and Practice", Maclaren and Son, London 1969, p 29.
4. Derjaguin, B.V. and Smilga, V.P., "Adhesion Fundamentals and Practice" Maclaren and Son, London 1969, p 152.
5. Furneaux, R.C., Thompson, G.E. and Wood, G.C., Corrosion Science,18, 1978, p 853.
6. Aircraft Process Specification, DTD 915B, Ministry of Supply, 1956.
7. Defence Specification, DEF-151, "Anodizing of Aluminium and Aluminium Alloys", Ministry of Defence, 1965.
8. Boeing Process Specification, BAC 5555, "Phosphoric Acid Anodizing of Aluminium for Structural Bonding", 1974.
9. Marceau, J.A., Firminhac, R.H., Moji, Y., "Method for providing Environmentally stable Aluminium Surfaces for Adhesive Bonding and Product Produced", USP 4,085,012, 1978.
10. Weber, K.E. and Johnston, G.K., SAMPE Quarterly, 1974, p 16.
11. Thompson, G.E., Wood, G.C. and Hutchings, R., Transactions Metal Finish., 58,1980, p 21.
12. Smith, T.V., MSc Dissertation University of Manchester Institute of Science and Technology,, 1983.
13. Bishopp, J.A., International Journal of Adhesion and Adhesives, 4 1984, p 153.